EXPLORATORY AND DESCRIPTIVE STATISTICS

THE SAGE QUANTITATIVE RESEARCH KIT

Exploratory and Descriptive Statistics by *Julie Scott Jones* & *John Goldring* is the 2nd volume in *The SAGE Quantitative Research Kit*. This book can be used together with the other titles in the *Kit* as a comprehensive guide to the process of doing quantitative research, but is equally valuable on its own as a practical introduction to Exploratory and Descriptive Statistics.

Editors of The SAGE Quantitative Research Kit:

Malcolm Williams – *Cardiff University, UK*

Richard D. Wiggins – *UCL Social Research Institute, UK*

D. Betsy McCoach – *University of Connecticut, USA*

Founding editor:

The late W. Paul Vogt – *Illinois State University, USA*

EXPLORATORY AND DESCRIPTIVE STATISTICS

JULIE SCOTT JONES
JOHN GOLDRING

Los Angeles | London | New Delhi
Singapore | Washington DC | Melbourne

THE SAGE QUANTITATIVE RESEARCH KIT

Los Angeles | London | New Delhi
Singapore | Washington DC | Melbourne

SAGE Publications Ltd
1 Oliver's Yard
55 City Road
London EC1Y 1SP

SAGE Publications Inc.
2455 Teller Road
Thousand Oaks, California 91320

SAGE Publications India Pvt Ltd
B 1/I 1 Mohan Cooperative Industrial Area
Mathura Road
New Delhi 110 044

SAGE Publications Asia-Pacific Pte Ltd
3 Church Street
#10-04 Samsung Hub
Singapore 049483

Editor: Jai Seaman
Assistant editor: Charlotte Bush
Production editor: Manmeet Kaur Tura
Copyeditor: QuADS Prepress Pvt Ltd
Proofreader: Derek Markham
Indexer: Cathryn Pritchard
Marketing manager: Susheel Gokarakonda
Cover design: Shaun Mercier
Typeset by: C&M Digitals (P) Ltd, Chennai, India

Library of Congress Control Number: 2020943060

British Library Cataloguing in Publication data

A catalogue record for this book is available from the British Library

ISBN 978-1-5264-2471-6

CONTENTS

List of Figures, Tables and Boxes   ix

About the Authors   xvii

1   Introducing Descriptive and Exploratory Statistics   **1**

What Is This Book About?   2

What's in Each Chapter?   2

New to Statistical Analysis? This Book Is for You!   4

So, What Are Descriptive Statistics?   4

What No Statistical Testing?   9

Sounds Like Inferential Statistics Are More Important   9

Types of Descriptive Statistics   10

    Categorical Data: Frequency Distributions   10

    Continuous or Interval-Level Data: Measures of Central Tendency   13

    Continuous or Interval-Level Data: Measures of Dispersion   13

One Variable or Two?   15

So, What Can I Do With Descriptive Statistics?   16

Why Not Try Exploratory Data Analysis   20

2   Finding Data to Describe   **23**

Introduction   24

Yours or Mine?   24

    What Is Primary Data?   24

    What Is Secondary Data?   26

Know Thy Data   27

    What Do We Mean by Research Data?   27

    Administrative Data   28

What's All This About 'Open Data'?   30

Show Me the Data!   32

    A Little Exploration First   32

    Data, Data, Everywhere   33

    Accessing and Downloading Data   34

    Are You Sure About This Data?   35

    Should I Go Large?   36

Downloading Data   37
   The General Social Survey   37
   The National Survey of Sexual Attitudes and
   Lifestyles 2010–2012 (Natsal-3)   42
Got Data: Let's Look Inside!   48
   Problem Buster!   49
Shit In, Shit Out: And Other Key Principles of
Data Management   49
Some Basic Cleaning Tips   50
Looking Ahead   53

3   Measure Everything – Learn Something – Answer Nothing:
An Exploration Into Variables and Types of Measurement   **55**

Introduction   56
Measurement as a Taken for Granted   57
Measuring the Social World   60
   Controversial and Contested Measurements   64
Units of Measurement: Variables   64
   Classification – Gender: Men and Women?   65
   Counting Gender: How Many Men and Women?   66
   Measurement: Different Experiences of Men and Women?   70
   Explaining: Different Experiences of Men and Women?   71
Levels of Measurement   72
Categorical Variables   74
   Nominal   74
   Ordinal   74
Interval Variables   74
   Ordinal   74
   Scale   75
Looking Ahead   77

4   I Am Not a Number, I Am a Categorical Variable   **81**

Introduction   82
Percentages: A Story of Parts and Wholes   85
Categorical Data and Percentages   87
Working With Valid Percent and Percent: Working
With Missing Data   95
   Presenting Percentages, Don't Forget the n   97
   What's This Cumulative Percent Column All About?   98
   To Merge or Not to Merge Responses   98

Writing Up Results: Producing Descriptive Summaries 99
  Show, Compare or Present 100
  Rounding Up or Down? 101
What to Report With Missing Data? 101
Strengths of Using the Percent 103
Working With Contingency Tables 105
How to Guides for IBM SPSS and MS Excel 107
How to Guide for IBM SPSS 107
  Carrying Out Univariate Analysis 108
  IBM SPSS Outputs 109
  Two's Company: Carrying Out Bivariate Analysis Using IBM SPSS 110
  Bivariate Analysis: Including Percentages 113
How to Guide for MS Excel 115
  Using MS Word to Make Frequency Tables 117
  Two's Company in MS Excel 124
Looking Ahead 127

5 I Like Being Average, I Am an Interval Variable **131**

Introduction 132
Measures of Central Tendency and Dispersion 132
Averages in Everyday Life: Feedback Scores 133
The Importance of Averages: Exploring Income and Poverty 134
Rich Man, Poor Man, Beggar Man, Thief: The Problem When
Measuring Average Incomes 136
What's Wrong With Being Average: Income Inequalities and the
Problem With Outliers 137
Measures of Dispersion: Adding More Context to the Data 141
Singing Stats! 143
  Standard Deviation: Horrible Name, but Really Useful 143
  Home on the Range? Using the Range Rule for a 'Quick'
  Standard Deviation 144
Average UK Income: But Which Average? 145
Bivariate Analysis Using MCT 147
When Averages Are Just Plain Silly 148
  'I Am Above Average Actually!' 149
Writing Up the Results: The End Is Nigh 149
'How to' Calculate MCT and MoD Using IBM SPSS 151
'How to' Calculate MCT and MoD Using MS Excel 155
Looking Ahead 155

6   Visualising Our Data   **159**

Does a Picture Tell a Thousand Words?   160
Does It Really Matter How I Present My Data?   160
There's a Graph for That   164
  The One With the Slices   164
  The One With the Bars   169
  The One With the Lines   174
  The One With That Sounds Like Instagram   176
  The One With the Whiskers   182
  The One With All the Dots   185
Spoilt for Choice!   189
What Makes a Good Graph?   189
Data Mapping   191
Looking Ahead   197

7   The Story Waiting to Be Told   **199**

Introduction   200
The Opposite Sex?   201
The Gender Pay Gap   202
Data Still Matters   203
Measurement Still Matters   205
'Let's Talk About Sex Baby'   205
The Importance of Context When Exploring Data   207
Beyond the Double Standard: Telling Stories of Sexual Difference 209
Men Estimate – Women Count: A Gender Difference to Recollecting 210
Trigger Warning! Statistics in the Media   210
The Transient Nature of the News Media   213
Butter Is Good/Bad/Good/Bad/WTF for You!   213
The Fine Line Between Mistakes and Misleading   216
The Devil's in the Detail!   217
Purveyors of Fake News!   217

Glossary   221
References   227
Index   233

LIST OF FIGURES, TABLES AND BOXES

List of figures

1.1	Population (*N*)	6
1.2	A sample from a population	6
1.3	Difference between a sample and a population	8
1.4	Prevalence of obesity by sex: Year 6 – NCMP data	19
2.1	Accessing the 'Get the Data' tab on the GSS website	37
2.2	The 'Get the Data' screen on the GSS website	38
2.3	Selecting data on the GSS website	38
2.4	Selecting GSS Explorer on the GSS website	39
2.5	GSS Explorer on the GSS website	39
2.6	'View Trends' button on GSS Explorer on the GSS website	40
2.7	'Is it wrong for same-sex adults to have sexual relations' by age, GSS Explorer	40
2.8	Accessing variables on GSS Explorer	41
2.9	Using keyword search on GSS Explorer	41
2.10	Results of keyword search on GSS Explorer	42
2.11	The Actions button on GSS Explorer	42
2.12	The UK Data Service website	43
2.13	The 'Details' tab for Natsal-3 on the UKDS	44
2.14	The 'Documentation' tab for Natsal-3 on the UKDS	45
2.15	The 'Resource' tab for Natsal-3 on the UKDS	46
2.16	The 'Access data' button for Natsal-3 on the UKDS	46
2.17	Data sets listed within a project on the UKDS	47
2.18	The Download screen on the UKDS	47
2.19	The downloaded Natsal-3 folders	47
2.20	periods2, Variable View, IBM SPSS	51
2.21	'cleaned' periods2, Variable View, IBM SPSS	51
3.1	Container ship	58
3.2	Double-decker bus	59
3.3	Full-time gross weekly earnings by sex	70

3.4	Full-time gross weekly earnings by sex and age group	71
3.5	Levels of measurement	73
4.1	Types of variables	82
4.2	'Raw' data taken from the Ice cream data set 2019 (IBM SPSS Data View, top)	87
4.3	'Raw' data taken from the Ice cream data set 2019 (IBM SPSS Data View, bottom)	88
4.4	Information for each variable	89
4.5	'Raw' data taken from the Ice cream data set 2019 (IBM SPSS Data View, text)	90
4.6	'Raw' data taken from the Ice cream data set 2019 (IBM SPSS Data View, showing text – toggle button)	91
4.7	Year on year comparisons of recorded homicide using knife or sharp instrument – frequencies	103
4.8	Year on year comparisons of recorded homicide using knife or sharp instrument versus assault – frequencies	104
4.9	Year on year comparisons of recorded homicide using knife or sharp instrument versus assault – percentages	105
4.10	Using Analyze on the IBM SPSS toolbar to access frequencies	108
4.11	The Frequencies dialog box in IBM SPSS	109
4.12	Selecting a variable in the Frequencies dialog box in IBM SPSS	109
4.13	Using Analyze on the IBM SPSS toolbar to access crosstabs	111
4.14	The Crosstabs dialog box in IBM SPSS	112
4.15	Selecting variables in the Crosstabs dialog box in IBM SPSS	112
4.16	The Crosstabs: Cell Display dialog box in IBM SPSS	114
4.17	Selecting percentages and counts in the Crosstabs (Cell Display) dialog box in IBM SPSS	114
4.18	Ice cream data set 2019 in MS Excel (top view)	117
4.19	Ice cream data set 2019 in MS Excel (bottom view)	118
4.20	Selecting a variable in MS Excel	118
4.21	Using the Insert Tab in MS Excel	118
4.22	The Create Pivot Table dialog box in MS Excel	119
4.23	Blank Worksheet for a Pivot Table in MS Excel	119
4.24	The Pivot Table Fields dialog box in MS Excel	120
4.25	Pivot Table of sex (CSEW 2016–2017) in MS Excel	120
4.26	Pivot Table of sex (CSEW 2016–2017) in MS Excel, with (blank) removed	121
4.27	Creating a Copy of the Variable Count in MS Excel	121

4.28   Converting counts into percentages in MS Excel   122
4.29   Pivot Table of *hrprel3* (CSEW 2016–2017) in MS Excel   123
4.30   Using the Pivot Table function to create Crosstabs in MS Excel   125
4.31   Inserting percentages into a Crosstab Pivot Table in MS Excel   125

5.1   Median average hourly pay for men and women between the
       years 2011 and 2017   148
5.2   Accessing Frequencies using Analyze on the IBM SPSS toolbar   152
5.3   Selecting a variable in the Frequencies dialog box, IBM SPSS   152
5.4   The Statistics dialog box in Frequencies, IBM SPSS   153
5.5   MS Excel selecting data from B2:B35421   156

6.1   Clustered bar chart of rsex and sffsatis   162
6.2   Scatterplot of dage and mastnum   162
6.3   Pie chart for AFTERLIF   164
6.4   Pie chart for AGE, GSS 2018   165
6.5   Pie chart for dagegr, from Natsal-3   166
6.6   Pie chart for agrp, from Natsal-3   167
6.7   Pie chart for agrp, from Natsal-3   167
6.8   Mortality amongst British soldiers during the Crimean war   168
6.9   'Exploded' pie chart for agrp, from Natsal-3   168
6.10   '3D' pie chart for agrp, from Natsal-3   169
6.11   Simple bar chart showing nominal data (CONARMY, GSS 2018)   170
6.12   Simple bar chart showing ordinal data (NIHILISM, GSS 2018)   170
6.13   Horizontal bar chart showing ordinal data (AFRAIDOF, GSS 2018)   171
6.14   Grouped bar chart of DEMANDS and GENDER1 (GSS 2018)   172
6.15   Vertical and horizontal stacked bar charts showing ordinal data
       (RELPERSN and MARHOMO, GSS 2018)   173
6.16   Simple line chart of HOMOSEX, GSS 1973–2018   174
6.17   Simple line chart of NIHILISM, GSS 2018   175
6.18   Multiple line chart of HOMOSEX and SEX, GSS 1973–2018   175
6.19   Multiple line chart of HOMOSEX and AGE, GSS 1973–2018   176
6.20   Simple histogram of periods2 from Natsal-3   177
6.21   Simple histogram of periods2 from Natsal-3, with line of
       distribution included   178
6.22   Stacked histogram of dage and contrameth from Natsal-3   179
6.23   Frequency polygon of dage from Natsal-3   180
6.24   Histogram of dage from Natsal-3   181
6.25   Population pyramid of dage and contrameth from Natsal-3   181

6.26   1D boxplot of AGE, GSS 2018   182
6.27   Histogram of AGE, GSS 2018   183
6.28   Guide to 1D boxplot of AGE, GSS 2018   183
6.29   Simple boxplot of Age and FEFAM, GSS 2018   184
6.30   Clustered boxplot of Age, FEFAM and Gender, GSS 2018   185
6.31   Simple scatterplot, AGE and WEIGHT, GSS 2018   186
6.32   Grouped scatterplot, AGE, WEIGHT and GENDER   187
6.33   Matrix scatterplot, AGE, HRS1 and TVHOURS, GSS 2018   188
6.34   Pie chart showing student ice cream preference   190
6.35   Bar chart showing political party preference – Presidential Election   190
6.36   Bar chart (with baseline on y-axis) showing political party
       preference – Presidential Election   190
6.37   Dr John Snow's Cholera map of Soho, London   192
6.38   Detail of Dr John Snow's Cholera map of Soho, London   193
6.39   'Find Your Neighbourhood' on Police UK website   194
6.40   Crime map of the University district, Manchester,
       Police UK website   194
6.41   'Sub-area' of Crime map of the University district, Manchester,
       Police UK website   195
6.42   List of crimes, sub-area of Crime map of the University district,
       Manchester, Police UK website   195
6.43   Higher Education Qualifications map, Manchester, DataShine   196

7.1   Gender-based disparities found across the world   204
7.2   Average number of lifetime sexual partners of men and
       women in the UK aged between 16–44   206
7.3   The building a narrative cycle   208

List of tables

1.1   Frequency distribution of Year 1 social science students'
       favourite class   11
1.2   Frequency table for age on IBM SPSS   12
1.3   Children's reading scores   13
1.4   Measures of central tendency and dispersion for reading scores   14
1.5   Contingency table showing film genre preference by gender   15
1.6   Frequency distribution of ethnicity, IBM SPSS output   17
1.7   Frequency distribution of sexual identity, Natsal 2010, IBM
       SPSS output   18

2.1   MCT, periods2, IBM SPSS   51
2.2   MCT for winsorized periods2   52
2.3   MCT for 'trimmed' periods2   53

3.1   UK population between 1937 and 2014   67
3.2   Variables identified from James's story   77

4.1   IBM SPSS frequency table for sex taken from the CSEW 2016–2017   91
4.2   IBM SPSS output frequency table for yrsarea (How long respondents
      have lived in the area)   94
4.3   Frequency table for wcarstol (How worried about having car stolen)   96
4.4   Merging data to summarise categories   98
4.5   How worried about being physically attacked   102
4.6   IBM SPSS output table for 'How worried about having
      car stolen' by sex   106
4.7   IBM SPSS output table for 'How worried about having
      car stolen' by sex (including column percent)   106
4.8   IBM SPSS outputs for HRP religion (CSEW 2016–2017)   110
4.9   Crosstab for religion and sex   113
4.10 Crosstab for religion and sex including percentages   115
4.11 Crosstab for religion and sex showing column percentages   116
4.12 Count of sex showing valid percent   122
4.13 Table before percentage grand total   124
4.14 Table including percentage grand total   124
4.15 Making a table in MS Word to show frequencies and valid percent 126

5.1   Hours per week online shopping   138
5.2   Hours per week online shopping (with MCT)   140
5.3   Hours per week online shopping (with results from MCT)   140
5.4   Hours per week online shopping (with results from MCT and MoD)   142
5.5   Median average hourly pay for men and women between
      the years 2011 and 2017   147
5.6   IBS SPSS output showing the MCT and MoD for respondent's age   154
5.7   MS Excel formula needed to calculate MCT and MoD using
      CSEW 2016–2017   155

6.1   Cross-tabulation of rsex and sffsatis   161
6.2   Frequency table for AFTERLIF, GSS 2018   165
6.3   Cross-tabulation of GENDER1 and DEMANDS (GSS 2018)   171

6.4   Cross-tabulation of RELPERSN and MARHOMO   173
6.5   MCT for periods2 from Natsal-3   177
6.6   MCT of dage and contrameth from Natsal-3   179

List of boxes

1.1   2-Minute Recap!   7
1.2   Reflective Exercise   9
1.3   2-Minute Recap!   16
1.4   Pause for Thought   20
1.5   Pause for Thought   21

2.1   2-Minute Recap!   26
2.2   Pause for Thought   26
2.3   2-Minute Recap!   29
2.4   Pause for Thought: You're a National Statistic!   30
2.5   Reflective Exercise: 'Access All Areas'?   31
2.6   Time to Get Your Hands Dirty! Finding and Downloading Data Sets 48

3.1   Pause for Thought   63
3.2   Reflective Exercise   64
3.3   Reflective Exercise   69
3.4   Pause for Thought: Mind the Gap!   72
3.5   Time to Get Your Hands Dirty! Identifying Variable Types   76

4.1   2-Minute Recap!   86
4.2   2-Minute Recap!   86
4.3   2-Minute Recap!   92
4.4   2-Minute Recap!   93
4.5   Pause for Thought   97
4.6   Time to Get Your Hands Dirty! Interpreting Frequency Tables   102
4.7   2-Minute Recap!   127
4.8   Time to Get Your Hands Dirty! Interpreting Crosstabs   127

5.1   2-Minute Recap!   134
5.2   Pause for Thought   135
5.3   2-Minute Recap!   137
5.4   Reflective Exercise   137
5.5   2-Minute Recap!   145

5.6 Reflective Exercise   147
5.7 2-Minute Recap!   148
5.8 Pause for Thought   149
5.9 Reflective Exercise   151

6.1 Pause for Thought   163
6.2 Reflective Exercise   176
6.3 2-Minute Recap!   188
6.4 Get Your Hands Dirty! Creating Charts and Graphs   189
6.5 Time to Get Your Hands Dirty!   191
6.6 Time to Get Your Hands Dirty!   196

7.1 Pause for Thought   204
7.2 Pause for Thought   206
7.3 2-Minute Recap!   211
7.4 Pause for Thought   212
7.5 Reflective Exercise   216

ABOUT THE AUTHORS

Julie Scott Jones is the Head of the Department of Sociology at Manchester Metropolitan University and former Director of the Manchester Metropolitan University Q-Step Centre, which received £1.15 million in funding from the Nuffield Foundation, Economic and Social Research Council and Higher Education Funding Council for England. She joined Manchester Metropolitan University in 2003; since 2018, she has been the Head of the Department. Julie is the author of *Being the Chosen: Exploring a Christian Fundamentalist Worldview* (2010) and co-editor of *Ethnography in Social Science Practice* (2010). More recently, she has co-authored a number of journal articles on the pedagogy of quantitative methods teaching, based on her current research based on this field. She currently teaches both undergraduate and postgraduate research methods modules, with a particular interest in ethics and quantitative data analysis.

John Goldring is the Co-Director of the Q-Step Centre at Manchester Metropolitan University, one of 15 centres across the United Kingdom to receive funding to promote the development of quantitative methods teaching across the higher education sector. Joining Manchester Metropolitan University in 2004, his initial research and teaching focus was on men, masculinity and health. He started teaching statistical analysis in 2012, and he developed a narrative approach to working with numbers based on Freirean principles of raising critical consciousness and challenging social injustice. Teaching on research methods units at both undergraduate and postgraduate levels, he has also successfully supervised a number of PhD students through to completion. In addition to co-authoring a number of journal articles on pedagogic approaches to teaching statistics, he has written on ethnographies of men's health.

1

INTRODUCING DESCRIPTIVE AND EXPLORATORY STATISTICS

Chapter Overview

What is this book about? 2

What's in each chapter? 2

New to statistical analysis? this book is for you! 4

So, what are descriptive statistics? 4

What no statistical testing? 9

Sounds like inferential statistics are more important 9

Types of descriptive statistics 10

One variable or two? 15

So, what can I do with descriptive statistics? 16

Why not try exploratory data analysis 20

Further Reading 22

What is this book about?

This book focuses on the uses (and some abuses) of what are called 'descriptive and exploratory statistics' and can be read as a stand-alone book on this subject. However, it is one of a series within the *SAGE Quantitative Methods Toolkit*, and there will be references throughout to other volumes in the series that cover relevant themes in greater depth than this book does. This volume (Number 3 in the series) builds on Volumes 1 (*Beginning Quantitative Research*) and 2 (*Survey Research and Sampling*), which you may find useful to read prior to this one, although there will be brief recaps of key material within this book. Likewise, Volumes 5 (*Archival and Secondary Data Analysis*) and 8 (*Inferential Statistics*) of this series develop further themes explored in this book, so you may find it useful to read these to develop your knowledge even further.

This book has seven chapters, each of which explores a key element of descriptive and exploratory statistics. We use IBM SPSS and MS Excel throughout and provide handy 'how-to' guides and accompanying screenshots to show how each software package does all the heavy lifting, meaning you don't have to. If you like to read books in a linear way, then each chapter of this book obviously builds on and links to the next. If you are a complete newcomer to this topic, then you should certainly do this, otherwise you may get confused; you will find that your confidence with the topic (and knowledge) will build with each chapter. However, if you already have some knowledge and understanding of the topic, you may wish to 'dip' into specific chapters.

What's in each chapter?

Chapter 1 discusses the definitions of and the differences between descriptive, exploratory and inferential statistics. It then explores why and how we can use descriptive and exploratory statistics both in their own right and as a precursor to inferential statistics. Chapter 1 should be your starting point if you are reading this paragraph and thinking 'I haven't a clue what descriptive, inferential or exploratory statistics means!'

Chapter 2 outlines how to access data from which to conduct statistical analysis, identifying both primary and secondary sources, before discussing some key principles of data management. This is the chapter for you if you have a research paper to do that involves accessing, cleaning and manipulating quantitative data, and you have never done it before. If you are clueless as to how to find good sources of data, then this chapter will help. On the other hand, it may be that you already have a good source of data but it is in the 'wrong' software format, or you feel overwhelmed by the fact that the data set has more than 10,000 cases and 700 variables. Don't worry, Chapter 2 will help you. Chapter 3 identifies the different ways we can choose to measure social concepts (which is a fundamental challenge for all researchers) and the consequences (good and bad) of those measurement choices. It introduces the concept of 'variables' and units of measurement. It is easy for students to get confused with different types of variables and to struggle with the concept of measurement; you may have already had a class on interval versus nominal variables or continuous versus categorical-level data and come out of it none the wiser. The whole thing may seem like a parade of words with little real meaning to you; try Chapter 3 before you decide never to attend another stats class, it will help. Chapters 4 and 5 build on Chapter 3, so don't start with them if you have not covered measurement before. Chapter 4 discusses categorical data, both nominal and ordinal; it defines, classifies and illustrates this type of data. It shows you how to analyse such data and acknowledge its limitations. There are many examples to help you understand. Chapter 5 does a similar thing with interval-level or continuous data. If you have done a bit of stats already, you may just dip into these chapters to refresh or reinforce your understanding; however, if you are like many students and struggle to understand the difference between ordinal-level data that is categorical and interval-level data that is continuous, then you may want to read both the chapters. It is worth noting that often students under-stand the differences between types of measurement but struggle to understand the greater significance of these differences; that is, why do the differences matter: if this is you, then take your time and read these chapters. Chapter 6 focuses on how to visualise our data, using different types of graphs. If you don't know the difference between pie, bar and stacked charts or struggle to understand why your tutor gets annoyed when you produce a colourful pie chart for the variable 'age' in your data, then this is the chapter for you. How you visualise your data is a skill in itself, and effective visualisations can communicate powerful stories about your data with less need for detailed discussion. Students often don't understand the power of good visualisation of data and how visualising your data effectively can assist the understanding and analysis of your data. Before you submit that research report with the same old frequency tables or poorly labelled pie charts, read this chapter. Chapter 7, the final chapter, discusses how to construct powerful narra-tives or 'stories' from your data. When your tutor asks you to discuss your findings at the end of your research report, you may think it is sufficient to summarise the key 'headlines' from your data analysis. Hopefully, you would support this discus-sion of analysis with some literature, but often students rush this element because they think the data findings themselves are the central focus. However, finding out that, for example, there is a rising trend on campus for plant-based diets among female students is not necessarily that interesting unless we can frame it into a wider narrative: is the trend common across a range of universities or indeed are women nationally participating in this trend? If it is a wider trend, is it to do with gender, age or educational level? Students often struggle to do this; when your tutor writes in her feedback 'You needed to develop this' or 'Why is this relevant?' she is identifying that you could have constructed a greater narrative for your data and it is time to try Chapter 7.

New to statistical analysis? this book is for you!

Statistics and quantitative data analysis books can be intimidating if you have not done any statistics or quantitative data analysis before; this book draws on material that has been tried and tested over the years on students who, perhaps like you, are a little bit nervous or even anxious about having to do statistical data analysis. This book is an introduction and an overview to the topic of descriptive and exploratory statistics; it will cover the basics that you would need to conduct your own descriptive-level data analysis whether at undergraduate or postgraduate level. We are presuming that you are new to the topic or are looking for a text that will reinforce or clarify your learning. To support your learning, this book has a number of features, including the following:

* *2-Minute Recaps*, where you test yourself against the clock, supporting your knowledge building
* *Pause for Thought*, time to stop and think, reinforcing your learning
* *Reflective Exercises*, where you can apply your learning via set exercises
* *Get Your Hands Dirty*, time to boot up the computer and do some data work

These different features will help build your conceptual knowledge whilst providing you with opportunities to practice your skills; practice is a key way that we can grow our confidence and understanding of key concepts through their application.

So, what are descriptive statistics?

A key distinction made in the field of quantitative data analysis or statistics is between descriptive and inferential statistics. To understand the difference between the two, we need to think about the basic goals of the quantitative approach; Volume 1 (*Beginning Quantitative Research*) of this series discusses this in much more detail if you are interested. Generally speaking, the quantitative approach draws on the scientific model of understanding the world; scientists attempt to 'draw' conclusions and understandings about the physical world through the development of hypotheses which they test using a range of experimental techniques. This testing leads to the development of theories or 'laws' about the physical world – for example, the laws of gravity or motion. The foundational element of this approach is that the physical world is knowable and understandable in an objective and quantified way. Social scientists use quantitative methods in a similar way to understand the social world, whether using experimental techniques (see Volume 4, *Experimental Design*) or social surveys (see Volume 2, *Survey Research and Sampling*).

However, in order to test hypotheses or answer research questions, social scientists first need to identify a population that they are interested in. A **population** refers to all the members of a specific group that we want to examine: the entire population of a country, all the students at a specific university, everyone aged 10 years old in a specific school district and so forth. Populations can be large, such as the entire population of a nation (which will be in the millions or even more than a billion) or small, such as all the schoolchildren in one town (possibly 500+) or even in one single school class (maybe 30 in total). Whoever our population are, they are the focus of our interest and study; we want to use quantitative methods (whether experiments or questionnaire-based surveys) to learn about them and possibly develop some social 'laws' or theories of our own. Some populations are so small that we can study them as a whole and generate **parameters**, which is data about a whole population. For example, we could, technically, gather data on all the students on a specific university course and generate parameters about them. To illustrate, we could gather data on all the 200 students enrolled on a single Criminology programme in one university; we could identify parameters for them such as average age, grades and so forth. The national censuses that are very common around the world try to do this for entire populations at the householder level; China's 2010 population census was the largest census ever conducted in the world, issued to more than 40 million households and gathering data on more than 1 billion people. Generally though, we acknowledge that it is technically and practically challenging to gather data on an entire population (see Volume 2, *Survey Research and Sampling,* for more detail on this), which is why national censuses tend to be on a 10-year cycle. Moreover, even amongst a small population of say 200 students on a course, we may struggle to reach them all, some may be absent from class or ignore emails and so forth, thus not completing our questionnaire. To overcome this problem, we tend to focus on samples. A **sample** is a part of a population, for example, there are 4.9 million mothers with dependent children (Office for National Statistics [ONS], 2017a) in the UK working either full-time or part-time. This is a large population, so we would draw a sample from it in order to understand the population better. If you are struggling to visualise this, think of a cake; the whole cake is the population, a slice is the sample.

Figure 1.1 represents a population that we might want to know more about. However, it is large, so we take a sample (*n*) from it that we can examine as shown in Figure 1.2.

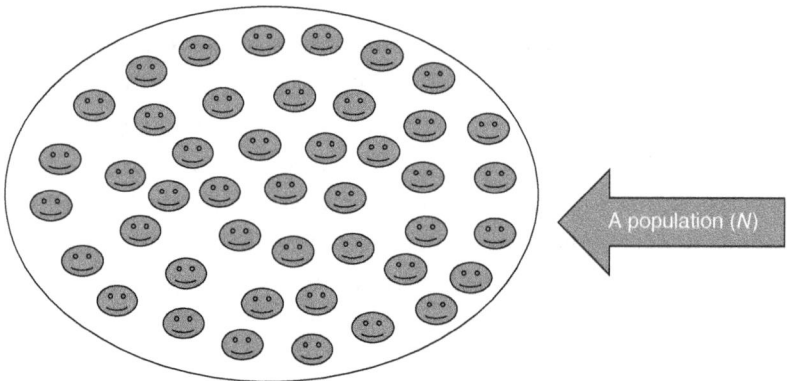

Figure 1.1 Population (*N*)

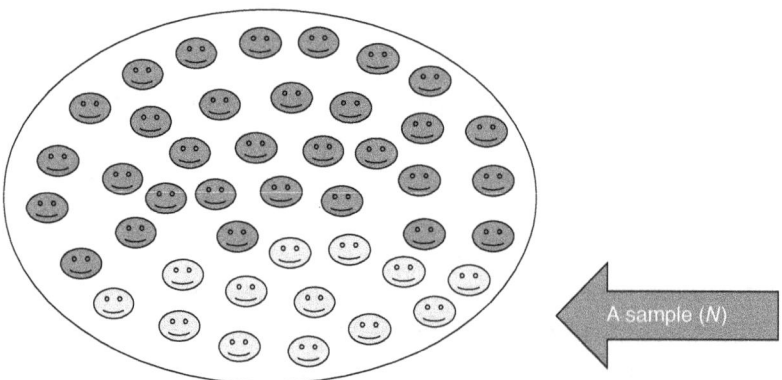

Figure 1.2 A sample from a population

The lighter shaded faces are our sample (*n*) taken from our population (*N*). We would generate descriptive statistics about the sample and then if our data was deemed sufficiently robust, we may be able to generalise from this sample back to our population.

Box 1.1

2-Minute Recap!

Set the timer on your phone and pair up with a classmate. Try to explain the following to your classmate who can check whether you are right (or not):

What is the difference between a sample and a population?

What is the relationship between a sample and a population in statistical analysis?

You might want to draw the difference and relationship if it is easier, especially if you are a visual learner.

Of course, a lot of this depends on the actual design of our survey tool and how our data is sampled. Volume 2 (*Survey Research and Sampling*) provides more detail about **sampling** strategies, but we can summarise the key issues as follows:

- *Sample size.* Typically, the larger the sample the more likely we are to be able to identify trends and patterns that may be important. If we only sampled 20 working mothers in the UK, it would seem ridiculous given the total population is 4.9 million. Common sense tells us that 20 working mothers could not possibly be representative of a much larger population. Of course, if there were only 30 working mothers in the whole of the UK, then 20 would be absolutely fine as a sample size. But why are small sample sizes problematic? If our sample is too small, then we are more likely to miss key trends or patterns within the broader population; likewise, we are more likely to overemphasise idiosyncratic trends unique to only a small group. The larger our sample the better, especially when we want to conduct inferential statistical testing, because most statistics tests rely on large samples (see Volume 8, *Inferential Statistics*, for a more detailed discussion).
- *Representativeness.* Samples need to represent our overall population, so we know that within the broad category 'working mother', there will be appropriate variation by age of child, number of children, relationship status, age of woman, type of work and working pattern, among many other potential variations. Therefore, we must ensure as much as we can that our sample represents the wider population. If we only sampled working fathers, non-working mothers or even just working mothers of teenagers, we would not generate representative data.
- *Randomly selected.* Randomness is crucial to the underlying philosophy of the quantitative approach. The social world is considered so complex that the chances of actually identifying key patterns or 'laws' are deemed low. To test this theory, we use data that has been randomly collected where every member of a population has an equal chance of being selected (using what is known as probability sampling, there are many different ways to do this). This means that

if we do detect, via testing, anything statistically significant, then it may very well be important. This links to the idea of generalisability where we seek to generalise about the population from the sample.

Figure 1.3 visualises the relationship between a sample and a population. In the social sciences, we have a way of showing if we have recruited the whole population or just part of it by using the capital '*N*' to show it is the population or the lowercase '*n*' (referring to the sample or a subgroup of the population). To return to our working mothers example, *N* would be the overall population of 4.9 million and *n* would represent our actual sample.

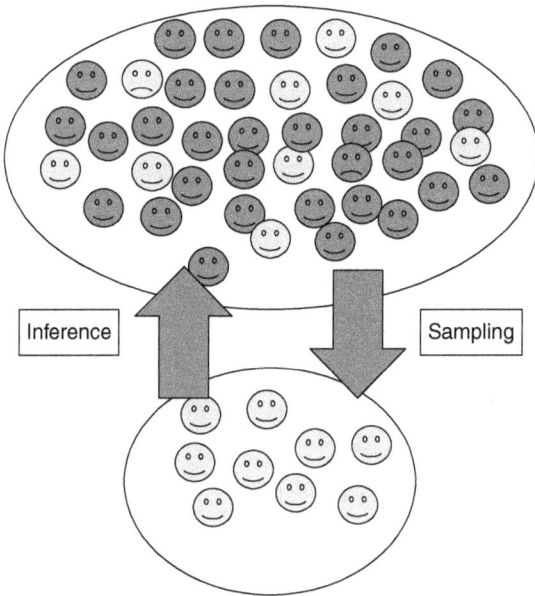

Figure 1.3 Difference between a sample and a population

At this point, you may be thinking what does any of this have to do with descriptive statistics, which was supposed to be the focus of this section? Well, descriptive statistics is when the researcher focuses on a sample (*n*) and refers to the statistics generated from that sample. **Descriptive statistics** allow us to describe, visualise (e.g. using graphs) and summarise the key features of our sample. We may have collected the sample data ourselves or accessed it from a secondary source, such as an online survey; nevertheless, we know who is definitely in our sample so that the analysis we generate is straightforward and (if we do it correctly) accurately describes this specific group. However, we cannot make inferences or generalisations about our population from our sample using descriptive statistics.

Box 1.2

Reflective Exercise

Imagine that you worked for the Department of Health, and your job was to analyse health data on your country's population. Let's imagine that your boss wanted to know more about the prevalence of poor mental well-being amongst the under 35s. What would your sampling strategy be? How would you ensure that your data was representative, random and large? What challenges might you face accessing this data? Can you think of ways to resolve these challenges?

What no statistical testing?

This may be the point when you ask, 'What about equations and funny-sounding stats tests? This descriptive stuff sounds too straightforward, can it be that easy?' Statistical testing in the social sciences is dominated by what is known as the frequentist or null hypothesis significance testing (NHST) approach, which involves formulating a hypothesis and then 'testing' it; the hypothesis in question is always the null. A hypothesis is a statement that you want to test in order to answer a research question. Here's an example of a (null) hypothesis: 'Male drivers are not more likely to get speeding fines than female drivers'. In our example, we are interested in whether there is a gender difference in the likelihood of collecting a speeding fine. The **frequentist approach** involves generating what are known as **inferential statistics** – that is, statistics from which we can make inferences about a population (N) from a sample (n). In this example, we want to make inferences about the behaviour of male and female drivers. This approach requires that the sample is obtained in the manner as outlined earlier in this chapter and in addition that the data meets a series of strict criteria known as parametric assumptions (see Volume 8, *Inferential Statistics*, for more information). What this means is that we cannot do statistical testing on a sample unless the data sample passes all these criteria or assumptions. A common mistake that students make is to try and test hypotheses from their descriptive statistics without testing if the data meets parametric assumptions.

Sounds like inferential statistics are more important

Although students may welcome the relative simplicity of descriptive statistics, they often feel or are made to feel (by classmates, peers or even tutors) that it is lesser or second best to inferential statistics. The dominance of the frequentist approach has meant that there is perhaps too much focus on inferential statistics without sufficient consideration of their limitations. Partly, this is also related to the fact that inferential testing can be more technically and conceptually complex and challenging than running descriptive-level analysis. Crucially, it is important to remember that each is attempting to do a different thing and that although the two may link within the research process (as we discuss in this chapter), descriptive statistics should be cherished in their own right, as a useful and potentially insightful form of data analysis. It is worth remembering that good description can be worth much more than poorly tested data; as this book will demonstrate, some of the most convincing and powerful narratives can be told using 'only' descriptive statistics. Furthermore, there is a growing school of thought that suggests that the frequentist approach with its focus on NHST is not appropriate and may cause greater misinterpretations of data than we think (see Ballinger, 2011; MacInnes, 2019b; White & Gorard, 2017). So next time someone sneeringly says, 'Are you *only* using descriptive statistics?' don't take the bait and rush to run loads of poorly chosen stats tests because you think it will impress your tutor more. Instead, cherish the simplicity (and challenge) of description.

Types of descriptive statistics

Once we have collected or downloaded our sample data, we have the challenge of trying to make sense of it. 'Raw' data cannot tell its own story, and with a large sample and/or many variables, we cannot expect the key trends or patterns to be clear to either others or ourselves. Therefore, it is important to generate a range of descriptive statistics and to learn what each type is telling us. The type of descriptive statistics we generate depends on the type of data that we have.

Categorical data: frequency distributions

Categorical data are variables, which are not numerical (unlike age, for example), and therefore, you cannot make calculations. Gender is a classic example of a **categorical variable**. A binary version of a gender variable would assign participants to either the 'male' or the 'female' category. We cannot calculate an average gender from this data (in the way we could calculate an average age), but we can generate frequency distributions to tell us about the distribution of males and females in our particular sample. Frequency distributions allow us to present the distribution of specific levels or categories within a variable across our sample. Typically, this involves creating a frequency table with relative and cumulative frequencies. Let's imagine that we collected data from 100 first-year social science students; we asked them what was their favourite class in their Year 1. In order to make sense of the data, we construct a frequency table, as shown in Table 1.1. We can easily see that 40% of our sample of 100 first-year social science students identify 'Introductory Statistics' as their favourite class. Likewise, we can see that the least popular course was 'Economic Theory' with only 10% of students selecting that as their favourite. A frequency table usually has four columns. The Frequency column is literally the number of times a specific data value occurs. The Percent column is the percentage of the frequency count for that specific data value out of the total sample number, whereas the Valid Percent column is the percentage of the frequency count for that specific data value out of the total number who answered that specific question. It is best to look at the valid percent as this is a more accurate percentage for a specific category or data value. The Cumulative Percent column adds the percentage of each category or level in the variable up to 100; it is useful for summarising majority responses.

Table 1.1   Frequency distribution of Year 1 social science students' favourite class

	Favourite Class: Year 1 Social Science Students			
	Frequency	Percent	Valid Percent	Cumulative Percent
Social policy in practice	38	38	38	38
Introductory statistics	40	40	40	78
Contemporary politics	12	12	12	90
Economic theory	10	10	10	100

For small samples, you can manually generate frequency tables, but for large samples, you will want to use software such as IBM SPSS, R or MS Excel. Chapter 4 provides a much more detailed discussion of how to generate and interpret frequency distributions, and Chapter 6 demonstrates how to visualise frequency distributions. It should be noted, particularly for the reader who may skip these chapters, you do not use frequency tables to summarise interval data. Table 1.2 shows a frequency table for a made-up variable *Age* that has been generated on IBM SPSS, using made-up data for the purposes of illustration only.

As Table 1.2 shows, the table is far too big to be able to make much meaningful sense. We might be able to state (using the cumulative percent) that 43% of the sample are under 30 and just over 25% are over 50, but it would be difficult to say much else. We certainly would not be able to calculate the mean or any other measures of central tendency (MCT) from this data in this format.

Table 1.2 Frequency table for age on IBM SPSS

Respondent's age at interview, years

Valid	Frequency	Percent	Valid Percent	Cumulative Percent
16	446	2.9	2.9	2.9
17	487	3.2	3.2	6.2
18	491	3.2	3.2	9.4
19	413	2.7	2.7	12.1
20	410	2.7	2.7	14.8
21	386	2.5	2.5	17.4
22	425	2.8	2.8	20.2
23	429	2.8	2.8	23.0
24	382	2.5	2.5	25.5
25	445	2.9	2.9	28.5
26	430	2.8	2.8	31.3
27	443	2.9	2.9	34.2
28	496	3.3	3.3	37.5
29	482	3.2	3.2	40.7
30	348	2.3	2.3	43.0
31	367	2.4	2.4	45.4
32	340	2.2	2.2	47.6
33	333	2.2	2.2	49.8
34	328	2.2	2.2	52.0
35	198	1.3	1.3	53.3
36	174	1.1	1.1	54.4
37	206	1.4	1.4	55.8
38	197	1.3	1.3	57.1
39	215	1.4	1.4	58.5
40	196	1.3	1.3	59.8
41	225	1.5	1.5	61.3
42	231	1.5	1.5	62.8
43	196	1.3	1.3	64.1
44	183	1.2	1.2	65.3
45	208	1.4	1.4	66.7
46	198	1.3	1.3	68.0
47	214	1.4	1.4	69.4
48	204	1.3	1.3	70.7
49	206	1.4	1.4	72.1
50	181	1.2	1.2	73.3
51	171	1.1	1.1	74.4
52	188	1.2	1.2	75.7
53	194	1.3	1.3	76.9
54	153	1.0	1.0	78.0
55	180	1.2	1.2	79.1
56	156	1.0	1.0	80.2
57	175	1.2	1.2	81.3
58	174	1.1	1.1	82.5
59	167	1.1	1.1	83.6
60	176	1.2	1.2	84.7
61	168	1.1	1.1	85.8
62	187	1.2	1.2	87.1
63	217	1.4	1.4	88.5
64	202	1.3	1.3	89.8
65	206	1.4	1.4	91.2
66	162	1.1	1.1	92.3
67	198	1.3	1.3	93.6
68	160	1.1	1.1	94.6
69	172	1.1	1.1	95.8
70	136	.9	.9	96.7
71	137	.9	.9	97.6
72	141	.9	.9	98.5
73	124	.8	.8	99.3
74	105	.7	.7	100.0
Total	15162	100.0	100.0	

We have a greater range of tools to generate descriptive statistics for continuous or interval data, which is numeric – for example, age, weight, BMI (body mass index), income and temperature. Again, we can manually generate these statistics, but it is easier to use software such as IBM SPSS, R or MS Excel to do it for you when you have a large sample. Chapter 5 shows you how to do this. Just as we did for categorical data, we want to know the distribution of our interval-level data across our sample; the two main ways to do this are as follows.

Continuous or interval-level data: measures of central tendency

A measure of central tendency is a value that identifies the central position or middle of our data; we can also refer to them as averages. These are the same averages that you learnt in primary school – the mean, median and mode. The mean (actually the arithmetic mean) is the total (or sum) of the value of each observation in your data divided by the total number of observations. The median is the middle value in your data when the values are arranged in an ascending or descending order. The mode is the most frequently occurring value in your data's distribution. Let's imagine that 20 children have had a reading test and we have collected their test scores. Table 1.3 shows the scores, which we have put into ascending order. How can we make sense of this data? It would be useful for us to know what the 'average' score is for this set of children.

Table 1.3 Children's reading scores

25, 32, 45, 48, 48, 52, 55, 55, 55, 58, 62, 62, 65, 65, 68, 70, 70, 72, 85, 90

The mean reading score is 59.1, which we can round down to 59. The median reading score is 60 and the mode is 55. It is particularly useful to compare the mean and the median because when data is approximately evenly (or normally) distributed these should be very close together or even identical; in our example, they are 59 and 60. The mean is very easily skewed by high and low scores (which is why comparing it with the median is always a good idea); in our data, we have a couple of very high and low scores, which may be why the mean is 59 while the median is 60. We might want to identify 60 as the most accurate representation of the average reading score for this data. The mode is a much more limited representation as we can see the mode is 55 but this is not the middle value of our data, and thus, we would not use the mode for this example. The mode should be used with caution, but it can be useful when looking at the distribution of categorical data; if our example had been looking at the children's shoe sizes, then the mode would be a useful average to use, as it would identify the most common shoe size in the sample. It is worth noting that you cannot run **measures of central tendency** for categorical data – for example, gender, social class or ethnicity – because this data is not numeric.

Continuous or interval-level data: measures of dispersion

MCT are nearly always used alongside **measures of dispersion** (MoD), which tell us about the spread of our data and its variability. There are a variety of MoD (all of which you can choose to calculate manually or by using statistical software like IBM SPSS or MS Excel). The easiest to understand is the range, which is the difference between our maximum value and our minimum value. For the data in Table 1.3, the maximum reading score is 90 and the minimum is 25; so, the range is 65. The size of the range tells us how spread out the scores are; in this case, our reading scores are well spread out, whether this is a good or a bad thing would require further examination of the data. We can also choose to divide our data into Quartiles or quarters: the first quartile (Q1) for our reading scores lies between the fifth and sixth students' marks (49); the second quartile (Q2) between the 10th and the 11th students' marks (60) and the third quartile (Q3) between the 15th and 16th students' marks (69). We can also calculate the interquartile range by subtracting Q1 from Q3 = 20. Quartiles are not as easily influenced by outliers and thus can be useful as a means to describe and understand the data. We can calculate the standard deviation to assess the spread of scores within our data and the variance to assess the data's variability; this can be either done using statistical software or manually using the standard deviation and variance equations. As a general rule, the smaller the standard deviation, the closer the scores are to the mean. A large standard deviation means that our scores are further from the mean and can often indicate that the data is skewed in some way or may just mean that there is greater variability in the data. The larger the variance, generally as a rule, the more our scores are spread out and variable; the smaller the variance, the more our scores are clustered towards the mean. Table 1.4 summarises all our MCT and MoD for our reading test scores.

Table 1.4 Measures of central tendency and dispersion for reading scores

n Valid	
Missing	**20**
Mean	59.1
Median	60
Mode	55
Standard deviation	15.63
Variance	244.2
Range	65
Minimum	25
Maximum	90
Percentiles	
25	49
50	60
75	69

The standard deviation for our data is 15.63, which is low, suggesting that we can 'trust' the mean as a measure of central tendency. The variance for our data is 244.2, which is large, suggesting the scores are spread out and variable.

One variable or two?

So far, we have been examining data by generating descriptive statistics for one variable, which is univariate (literally meaning one variable) analysis. We can also examine the relationship between two variables; this is bivariate (literally meaning two variables) analysis. We can generate **contingency tables** (also known as **cross-tabulations or crosstabs** for short) for categorical data. Let's imagine we wanted to explore whether there was a gender difference in film genre choice, and we collected some data asking 50 men and 50 women what their favourite film genre was. Table 1.5 shows the results.

Table 1.5   Contingency table showing film genre preference by gender

Favourite Film Genre	Gender		Total
	Male	Female	
Sci Fi	15	5	20
Rom Com	20	5	25
Action	5	10	15
Horror	10	30	40
Total	50	50	

We can see from Table 1.5 that there seems to be a gender difference in film genre preference, with females identifying 'Horror' (30 out of 50) as their majority favourite, whereas males' majority preference is 'Rom Com' (20 out of 50), albeit closely followed by 'Sci Fi' (15 out of 50). We can also note that there is no genre that both genders share as a clear favourite. Because this is descriptive statistics, we cannot say that all men and women (in our overall population) have such film preferences, but for our sample, we can say that gender does appear to influence film genre preference. A contingency table therefore is a means to explore potential associations between two categorical variables. As Chapter 4 will demonstrate, you can add more content to contingency tables (including expected counts and percentages) either manually through doing your own calculations or you can choose to run them on statistical software. Likewise, in Chapter 5, you will learn how to generate and examine bivariate descriptives for the distribution of a **scale variable** across the levels or categories of a categorical variable, manually or via statistical software.

Box 1.3

2-Minute Recap!

Set the timer on your phone and pair up with a classmate. Try to list as many different types of ways that we can describe our data, and can you identify what type of data suits which type of descriptive?

So, what can I do with descriptive statistics?

So now that we have established what descriptive statistics is (and isn't), we can focus on the three main ways in which you could use them within your work. The three ways are presented as separate, but you should appreciate that they overlap and you may find yourself using descriptive statistics in all three ways.

1 Explore your data.

It may seem obvious (but not always to some) that before you can do anything with your data that is meaningful, you need to explore it. This means running descriptives of your 'raw' data ('raw' data being data that has not had any 'cleaning') using frequency tables and/or MCT and MoD. You might run some contingency tables too. The outputs will look a bit messy because the data is 'raw', but by taking the time to do this (and let's face it, if you use statistics software it wouldn't take much time at all), you will literally see what's in your data sample. This is often the moment when you realise that your sampling and/or data collection strategies did not quite go according to plan; for example, there may be one group overrepresented which may skew your data. Data exploration should be the first stage of any form of statistical work, whether it be for a report based solely on descriptive statistics or the foundation of a complex piece of inferential analysis. If you don't look 'inside' your data and see what's in there you will make mistakes at all other stages of your work. Statistics software makes it easy for us to run descriptives quickly but that can mean that we rush this stage, don't! Take your time. Look at the 'raw' data prior to running any outputs; if your sample is small, you will literally be able to review each case at a time, by doing this, you will see potential patterns, anomalies and gaps. If your data set is large, this is harder to do. Data exploration should be an active stage, don't just passively review your 'raw' outputs. Instead, ask questions of the outputs, such as 'what's the trend', 'who's missing', 'who's overrepresented', 'what's surprising', and 'what's expected'. By doing this, you are already starting to think about how you may clean your data (everyone has to do some amount of **data cleaning**) and how you might construct a narrative around this data.

Make notes against each output, which you can return to at a later point. If you don't take notes whilst it is fresh in your head, you may miss important observations.

2   Clean and prep your data.

Whether you are working with primary or secondary data, you should always pre-sume that you will have to do some cleaning to it. You cannot know what or how to clean your data unless you use descriptive statistics to help you; some students do try and clean data without actually reviewing frequency tables and such like, but this approach never ends well. Use descriptive statistics to help you make informed data cleaning choices. 'Raw' data is the data in the form that you collected or downloaded it; it is ground zero for your data. Anything that you choose to do to your data that changes it from its original 'raw' state is known as 'cleaning'. We clean data in order to conduct our analysis and there are two key functions of 'cleaning':

1   *Remove errors*. All data will contain errors, even national government sourced data. Most errors are input errors where someone has inputted the wrong value. If such errors are not removed, then they can make your output tables look untidy and can potentially skew your data. Errors are usually very easy to spot. If we look at Table 1.6, which shows the frequency distribution of the variable 'Ethnicity', we can see that there is a category 'male', which is clearly an error.

Table 1.6 Frequency distribution of ethnicity, IBM SPSS output

R's Ethnicity

		Frequency	Percent	Valid Percent	Cumulative Percent
Valid	White British	30424	89.1	89.2	89.2
	Dual-Heritage British	372	1.1	1.1	90.3
	Asian British	2078	6.1	6.1	96.4
	Black British	930	2.7	2.7	99.2
	Male	285	.8	.8	100.0
	Total	34089	99.8	100.0	
Missing	System	74	.2		
Total		34163	100.0		

2   *Preparation for analysis*. Once errors are removed, all other cleaning involves decisions that should be guided by what you need to do to your data to prepare it for analysis. Your preparation depends on what sort of analysis you want to do. If you are doing a descriptive level analysis, then you need to avoid over-cleaning your interval-level data regarding issues such as outliers and skew, because these can be insightful at this level of analysis. Obviously, if you are planning to go further and conduct inferential tests, then you will need to do quite a bit of cleaning to make your data meet parametric assumptions. The most significant consideration is theoretical and that means what you choose to do with specific variables, especially

those with many levels. This is particularly the case when working with secondary data, which someone else has collected. We illustrate with the variable 'Ethnicity'. In the UK, most national research uses the ONS' classification of ethnicity which has 18 categories. Your research may only be interested in identifying respondents as 'White' or 'BAME'. Therefore, you would need to clean this variable from 18 categories to two. How you choose to clean a variable should be based on theoretical considerations that you can justify typically based on a review of existing studies and literature on the topic. Table 1.7 shows the frequency distribution of sexual identity from the 2010 Natsal survey in the UK.

Table 1.7 Frequency distribution of sexual identity, Natsal 2010, IBM SPSS output

Sexual identity

		Frequency	Percent	Valid Percent	Cumulative Percent
Valid	Heterosexual/straight	14617	96.4	96.4	96.4
	Gay/lesbian	213	1.4	1.4	97.8
	Bisexual	226	1.5	1.5	99.3
	Other	53	.3	.3	99.7
	Not answered	53	.3	.3	100.0
	Total	15162	100.0	100.0	

Note. Natsal = National Survey of Sexual Attitudes and Lifestyles.

There are four categories how you might clean this data. If you were just theoretically only interested in comparing heterosexuals with non-heterosexuals, then you might recode the data into two groups. Obviously, if you were interested in different sexual identities, you might keep all the categories except 'Not answered'. However, what would you do with 'other'? Is it meaningful or not, might it hint at an emerging sexual identity? You also need to think about whether the 'Not answered' category is also meaningful, as it might hint at respondents' unease at answering the question. That said, only 53 did not answer out of 15,162, so that suggests most do not have an unease with the topic. These are all theoretical considerations, but these need to be weighed with practical ones. If you wanted to do some inferential testing, then you need to consider if three small groups should be retained as most tests struggle with small unequal levels in a variable. Thus, we must weigh up the theoretical considerations with the practical when we clean our data. Linked to this is the issue of narrative and what story you want your data to tell; this may also influence how you choose to clean your data.

Data cleaning again is often a stage that students rush in order to get to the analysis and is often tackled in an instrumental or passive way; in other words, you do it without thinking too much about it. It is worth remembering the central rule of data cleaning: 'shit in, shit out'. In other words, if you do not take the time to clean and organise your data, then your end result will not be good.

3   Tell a story with your data.

Descriptive statistics can provide you with a level of analysis in their own right: if you take the time to examine and clean your data carefully and present it clearly, framing your analysis with a good range of academic source materials, then you have the ingredients for a really powerful and easily accessible (because the reader doesn't have to plough through lots of technical outputs relating to testing) narrative. This is safe in the knowledge that you can make definitive statements about your sample and are not making inferences about a wider population, which may be incorrect given the increasing awareness of the potential deficiencies in the frequentist approach. A good example of a significant story, which can be told by mere descriptive statistics, is the increase in childhood obesity in the UK. The UK Government runs the National Child Measurement Programme (NCMP) that involves the measurement (weight and height taken, BMI calculated) of all children in their first and final years of primary school, roughly 1 million children per year. The NCMP started in 2006 and its data is freely available online (https://digital.nhs.uk/services/national-child-measurement-programme/). The programme also identifies school type, location and indicators of the child's socio-economic status. The graph in Figure 1.4 shows the prevalence of obesity amongst children in their final year (Year 6) at primary school. The data shows obesity by gender and across the first decade of the NCMP.

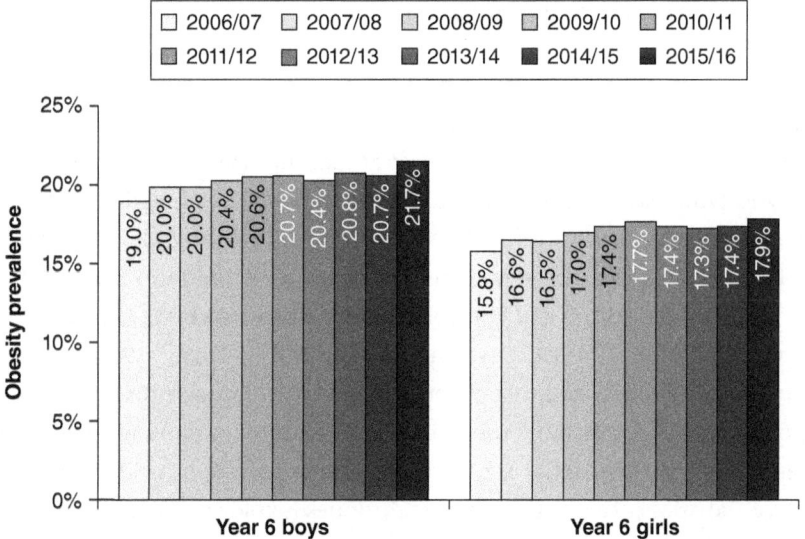

Figure 1.4   Prevalence of obesity by sex: Year 6 – NCMP data

Source. https://www.slideshare.net/PublicHealthEngland/patterns-and-trends-inchild-obesity-june-2017.

Note. NCMP = National Child Measurement Programme.

Figure 1.4 is a good example of descriptive statistics, showing a powerful narrative: the percentage of boys and girls in their final year at primary school in England who are obese is increasing with a slight tendency towards boys being more likely to be obese than girls. How might we account for this change amongst England's 10- to 11-year-olds and why boys? Well, we would need to do some further research on the topic, but we could probably construct an interesting narrative that includes the growth of social media and gaming, a decline in outdoor play and available play spaces, not to mention the rise of risk-aversive parenting (to name just a few potential elements) all of which contribute to increasingly sedentary lifestyles amongst young people, which can cause obesity. Look what we've done with just a few descriptive statistics (gender, BMI and year data collected).

Box 1.4

Pause for Thought

There are three ways to use descriptive statistics, can you remember each one and do you understand why each is helpful for your analysis?

Why not try exploratory data analysis

Those of you who are particularly observant may be wondering about exploratory statistics which was mentioned at the start of this chapter. So where do exploratory statistics fit in all this and what are they? Sometimes the term *exploratory statistics* is used synonymously with descriptive statistics; as we noted earlier, it is important to explore your data using descriptive statistics prior to any cleaning, analysis or further testing. However, the more precise meaning of the term *exploratory* in this context refers to something called **exploratory data analysis (EDA)**, which was first developed by the famous statistician John W. Tukey. Tukey (1970) was critical of the frequentist approach and instead encouraged an approach that was less focused on hypothesis testing and more on using a variety of techniques to explore the data for interesting and potentially significant patterns and trends. EDA overlaps with descriptive statistics because it uses the techniques (exploring data via frequency distributions, MCT and MoD) that we outlined earlier; indeed, Tukey championed the use of the maximum, minimum, median and quartiles. It should be noted that EDA also incorporates elements of data cleaning where so-called robust techniques are used to 'clean' data. EDA has become increasingly influential, as researchers grow critical of the frequentist approach and the use of EDA techniques (greater use of visualisations, a focus on different descriptive techniques, an interrogation of data quality and origin) has become more widespread. If you are interested in knowing more about EDA, then see MacInnes (2019a) for a good overview of the topic or review Tukey's own *Exploratory Data Analysis* (1970), which established the EDA approach. A key point to make is that you can use descriptive statistics to analyse your data without being committed to doing an EDA, but you may find yourself drawing on the influence of the EDA approach to really examine your data with the great care that Tukey and others sought. We can use some of the core questions that the EDA approach poses to examine our data and think through data cleaning:

1   How was this data gathered and for what purpose?
2   What kind of data do I have; what are the pros and cons of this type of data?
3   What do visualisations of my data tell me?
4   What is missing from my data and is it a problem? If it is a problem what can I do about it?
5   What errors are in my data and how can I remove them?
6   Are there outliers in my data and should I care about them?
7   What kind of question(s) do I want my data to inform or answer?
8   What can I add, change or remove from my data to influence my analysis and what are the consequences of these actions?
9   What are the key trends in my data and what do I make of them?

Box 1.5

Pause for Thought

In this chapter, you have learnt some key words and distinctions. Do you understand the basic difference between descriptive, exploratory and inferential statistics?

Chapter Summary

This chapter has been a bit of a whistle stop tour of key terms and concepts, so you may feel a little out of breath. Remember the subsequent chapters are at a slower pace and go into all these concepts and processes in much greater detail. The core takeaway from this chapter is that descriptive statistics are essential for any researcher irrespective of the type of analysis she or he is going to pursue. If you don't take the time to look inside your data and get to know it, then the 'shit in, shit out' rule will apply, and you will find yourself ruing the day when you rushed through this foundational work because you thought (wrongly) that your tutor would be more easily dazzled by more 'complex stats'.

Further Reading

Field, A. (2019). Discovering statistics with IBM SPSS. Sage.

This is a massive textbook that covers probably everything you ever wanted to know about doing statistical analysis. It is written in a funny, accessible style aimed at undergraduates. It is huge and covers a lot, so you might want to start by reading the first couple of chapters, which introduce key concepts and issues. There is a companion website which has some great study resources, including quizzes.

Kranzler, J. H. (2017). *Statistics for the terrified*. Rowman & Littlefield.

This book is useful if you want to study a little more in depth about descriptive statistics, especially Chapters 4 and 5. If you are, as the book title says, terrified by the thought of doing statistics, you might also look at Chapter 2, which looks at maths/stats anxiety and provides some practical solutions that could help get over this barrier to engaging with statistics. A small word of caution however, this book does work with formula.

2

FINDING DATA TO DESCRIBE

Chapter Overview

Introduction 24

Yours or mine? 24

Know thy data 27

What's all this about 'open data'? 30

Show me the data! 32

Downloading data 37

Got data: let's look inside! 48

Shit in, shit out: and other key principles of data management 49

Some basic cleaning tips 50

Looking ahead 53

Further Reading 54

Introduction

We expect that Chapter 1 has probably got you all excited about the possibilities of data analysis and you are eager to get stuck in. However, before the fun can really start, we need to find some data to analyse. This chapter is all about how to get your hands on data; you can read it as a stand-alone chapter, for those of you who know all about descriptive statistics but are clueless as to where to find data. You may have always been provided by nice and tidy data sets by your tutor(s), and now they are challenging you to find your own, perhaps for a dissertation or end-of-year research report. This can feel daunting but not to worry, this chapter will help you with this task. Alternatively, you may be reading this chapter in conjunction with the others; it makes sense to us as authors to start with finding data before we go deeper into different types of descriptive statistics. Your tutor may not expect you to find your own data just yet, but it is good to know what is 'out there' which you may use in the future. This chapter will look at three key elements: firstly, it will show you how to access data; secondly, it will discuss data management; and lastly, it will discuss some basics of data cleaning. Remember that this book is one of a series and the issues covered are also discussed in other volumes in the series, specifically Volumes 1 (*Beginning Quantitative Research*), 2 (*Survey Research and Sampling*) and 4 (*Archival and Secondary Data Analysis*).

Yours or mine?

The first decision we must make is whether we want to collect our own data (known as primary data collection) or use data collected by other researchers (known as secondary data collection). As we will see, there are pros and cons to using both types, which we must consider prior to starting any piece of research; the type we ultimately choose will shape our analysis, limiting or expanding its scope.

What is primary data?

Primary data is data directly collected by the researcher; primary quantitative data is usually gathered using questionnaire-based or attitudinal surveys (or indeed surveys which combine both elements). Volume 2 in this series (*Survey Research and Sampling*) covers the specifics of survey design and method in much more detail than this chapter does, so you might want to read it if you are planning to collect your own data. Researchers use a variety of ways to distribute surveys, including telephone, mail and increasingly online (using social media, websites and email).

Indeed, these days most of us are possibly quite used to completing short online surveys seeking to assess how 'satisfied' we may be with a particular product or service. The advantage of research using primary data is that researchers shape and control the whole process including design and distribution; therefore, they can be confident that the data collected best fits with their particular research question. This is often why students want to conduct their own quantitative data collection. Yet this is also a potential weakness of working with primary data: it is only as good as you (the researcher) make it. Can you guarantee that your attitudinal survey is valid, that is, it actually measures the attitudes you want to measure? And is it reliable? In other words, could someone else use your attitudinal survey and obtain similar results? Designing your own data collection puts much more pressure on you to get the design 'right', which takes more time (extensive literature reviews, running multiple pilot surveys and refining questions) than you might think. Surveys are notorious for their low response rates (as low as 10%–15% for some types), especially as people are increasingly bombarded with online or email customer surveys. It can take considerable resources and time to design-in good response rates. There is also the potential for over- or under-representation of specific groups, depending on your sampling strategy. When you start factoring in these sorts of issues, then primary survey-based quantitative data collection becomes much more costly in time, effort, data quality and possibly money than it first appears. It is not as simple as sending out a quick email questionnaire. The key strengths of quantitative methodology, as we noted in Chapter 1, is its generalisability from a sample to a large population; this requires large amounts of robust data. This is not to say that it is not good to collect your own data, it can be and it certainly allows you to understand (through application) key research design concepts, including sampling, measurement, validity and so forth. Rather, it is important to consider if it is the best approach for you and the research questions that you are trying to address, within possibly specific assessment time limits or demands. For example, if you were interested in what students in your university think about the quality of campus food, then you would have to conduct your own survey; it is unlikely that anyone else has collected this data. It might also be relatively easy to gather a large enough sample through a combination of paper-based questionnaires and email, particularly if you hung around the campus food outlets to gather respondents. It is also a topic that your respondents may have very strong views about, making them more likely to engage with your survey. However, if you wanted to explore respondents' views on the growth of plant-based diets among the under-30s, then it is unlikely that you would be able to collect data as good as that already collected by a variety of national (or even international) surveys, so then you may want to use secondary data instead.

What Is secondary data?

Secondary data is data collected by someone else for a specific purpose, such as an opinion poll or a government census. Working with secondary data is time- and cost-efficient, particularly in the digital age, where due to the commitment by researchers, governments and other organisations to 'Open Data', large amounts of data is literally at our fingertips. In addition, such data has already been ethically approved and has high levels of validity and reliability, making it a great resource for a student researcher. Typically, such data sets will be large, providing the student researcher with more data than she could ever have collected herself; moreover, that data will most likely be robust and allow for effective statistical testing. The disadvantages, of course, are that this is someone else's data, collected for a specific purpose, without your research project in mind. Therefore, you may find the wording of questions, type of data collected and so forth, not quite what you would like, and therefore, you will have to try your best to 'fit' the data to your research project. Volume 5 in this series, *Archival and Secondary Data Analysis*, provides more information about secondary data analysis – its pros and cons.

Box 2.1

2-Minute Recap!

Set a timer on your phone and list all the pros and cons of working with primary data – you've got 1 minute!

Then repeat this, but this time list all the pros and cons of working with secondary data – again you've got 1 minute!

You might want to do this in a pair so it becomes competitive.

Box 2.2

Pause for Thought

Consider the following research topics and decide whether you would use primary or secondary data collection to investigate them:

- Student consumption of plant-based diets
- Children's use of social media
- Women's attitudes to migration
- Men's consumption of grooming products
- Working parents and work–life balance

Know thy data

Obviously when you collect your own data, you know exactly what type it is and why you collected it. However, when working with secondary data, it is important to consider the reason why the data was collected in the first place. We can categorise secondary data into two broad categories: research data and administrative data.

What do we mean by research data?

Research data is data collected for the sole purpose of academic research, usually in order to answer set research questions and achieve specific objectives, such as creating a new product, testing a theory or evaluating a clinical intervention. Research data is collected by researchers within a variety of settings: academic (e.g. a university or a research centre), commercial/industrial (e.g. a factory or a product development lab) and public (e.g. government departments or agencies). Research data is usually highly robust and reliable, but questions are often raised regarding how 'reliable' data can be that is collected within specific funding contexts. For example, if you are a researcher paid to evaluate car emissions for a large car manufacturer, can you really be truly 'neutral' and unbiased? Likewise, if a government department is funding your evaluation of a very high-profile prison rehabilitation programme, then you may feel under tremendous pressure to 'water down' your results if you find the programme to be ineffective. Of course, this sort of behaviour is unethical, but these are exactly the pressures that researchers can face when doing funded research. Although research data is usually robust, we should appreciate the contexts within which researchers work, especially on funded research. It would be naive of us to presume that all research data is completely neutral and unbiased. That said, most governments fund large-scale attitudinal (in the UK there is the British Social Attitudes Survey (BSA); in the USA, the General Social Survey (GSS); and in the European Union (EU), the European Social Survey) and cohort studies as part of their commitment to national research programmes. In the UK for example, the government continues to fund a series of birth cohort studies (see https://www.closer.ac.uk, for more information), which have tracked different generations since the 1950s and have provided powerful insights into changing behaviours, trends and attitudes over the past 60 years. Such programmes are in the national interest, contribute to collective knowledge and allow researchers across a range of disciplines to analyse key national trends and phenomena, typically with policy impacts. An example of this impact on policy is the use of data from the BSA regarding attitudes to homosexuality in the UK. In 1983, the BSA asked respondents their views on homosexuality for the first time; 62% said that it was 'always/mostly wrong', but by 2003, this figure had fallen to 40%, whereas

41% said it was 'not wrong at all/rarely wrong'. In 2012, only 28% deemed it to be 'always/mostly wrong' against 47% who stated it is 'not wrong at all'. This is a major change in views over time. This data meant that the UK government could introduce laws that legally recognised same-sex relationships (civil partnerships in 2004 and same-sex marriages in 2013), knowing that the public's views had liberalised sufficiently to ensure that introducing these new laws was not controversial and more likely to be accepted.

Administrative data

Administrative data is data collected by a variety of organisations, most notably by governments, to maintain a record of key elements of their business or service delivery. Administrative data enables organisations to do a number of things:

- *Keep user records*. Organisations need to know who is using their services, whether it be a hospital, a private gym, or a university. Such 'census' data can allow organisations to refine services, target specific users, or simply be able to report usage. To illustrate, governments need to know the user rates of hospitals, prisons, schools and suchlike in order to financially budget and plan.
- *Identify key trends*. User data allows organisations to identify key patterns in their data and changing trends. The registration of births and deaths is a key way for governments to know population trends. If the birth rate is rising, then there will be a greater demand in the future for school places and perhaps new schools need to be built. Likewise, the declining fertility rate and increasing longevity of populations across the Western world are key trends, which raise important social policy challenges for governments in relation to migration, pensions, taxation and employment.
- *Deliver services*. Knowing how many users you have enables organisations to plan and budget for service delivery. A university, for example, needs to know how many students it has in order to ensure sufficient levels of service on campus. Relatedly the ageing population trend means that many Western governments need to plan for greater social care and pensions provision.
- *Evaluate performance*. Organisations can use administrative data to evaluate how effective their operations are; relatedly, governments can use such data to evaluate whether their services are being delivered effectively. For example, the results of national school tests in the UK are used to evaluate the effectiveness of schools and therefore educational policy; likewise, the reduction in the number of criminals caught could be used to evaluate the effectiveness of the police.
- *Keep people informed*. The publication of administrative data is seen as a means to ensure accountability in government and other organisations. The shift towards 'open' reporting of data is a means to 'democratise' data and enable citizens to have insight into how their taxes are being spent and the quality of the services delivered on their behalf. Of course, this relies on citizens being statistically literate and on how 'open' governments and other organisations are in what data they publish. It will not surprise you to know that what governments (and other

organisations for that matter) choose to publish is decided by them and not us; therefore, we should always give extra scrutiny to such data.

2-Minute Recap!

Set a timer on your phone and identify the key differences between research and administrative data – you've got 2 minutes!

   You might want to do this in a pair, with each of you taking turns to identify a difference, so it becomes more competitive.

Examples of administrative data

Administrative data is typically collected in two main ways; firstly via registration forms (completed when a user engages with a service), and secondly via census reporting. When a user wishes to access a specific service or legal status, he will typically complete a form, whether to register as a patient, a student or as 'unemployed' (to gain access to welfare). Sometimes registration is a legal requirement – for example, the registration of births, marriage and deaths. These forms are a source of administrative data that is then recorded, stored and reported. The reporting of such data may be via national offices of statistics, organisational websites and/or end-of-year reports. Census reporting is the collection of administrative data through the counting of activities with the aim of assessing trends, typically 'performance'. For example, when school test papers are collated and marked, the overall marks are 'counted' per school, district and nationally to provide an overall picture of pupil 'performance'. In the UK hospital, mortality data is collated and used to assess hospital 'performance'.

   National Censuses are good examples of large-scale administrative data; a large number of nations conduct these on-set? cycles. Censuses overlap both types of administrative data. In the UK, the Census runs every 10 years, and it is a legal obligation to complete the form; it is an important source of data for the government. Censuses provide 'snapshots' of populations at specific time periods, which when compared can show often dramatic changes in populations, such as the rise in single households or the changing ethnic composition of specific areas. Censuses can be used to corroborate other analyses of trends that have been done on an annual cycle; a trend towards fewer birth registrations over a couple of years may be dismissed as a 'blip' until compared with Census data on an overall population, which may show a more consistent overall decline.

Box 2.4

Pause for Thought: You're a National Statistic!

List as many ways in which the government (and its related agencies) has collected your data from your birth to the present day.

Do you ever think about how they store your data and who has access to it? Do you trust them to use your data wisely?

Can you trust administrative data?

One of the potential criticisms of administrative data is that it is not collected for research purposes but is literally 'counting' data to serve the purposes of specific, typically governmental organisations. Because governments and related agencies control how administrative data is collected, there is great potential for both bias and error. The issue of bias is a general issue that faces all researchers: how we ask our questions often shapes our results. Academic researchers typically try to remove bias and seek neutrality in their approach, although not always. Governments have perhaps more at stake, and therefore, how they choose to define or collect data is typically shaped by their policy aims. This is particularly the case with controversial social phenomena like unemployment, poverty and immigration. A government anxious to be seen to be reducing unemployment, crime, or poverty may actively change how they measure and record such data to produce a declining trend outcome. Therefore, it is important for us to be cautious in how we approach such data, and we must be very clear as to how it was measured and collected. Because administrative data is not collected for 'purely' research purposes, it can also be subject to higher levels of error and missing data as it often relies on non-researchers to collect it. A good example of this is UK Police data, which is collected via the recording of incidents by ordinary police officers, who may not have sufficient time to consistently or accurately record all elements required or may miss elements.

What's all this about 'open data'?

You may have noticed that this chapter has alluded to something called 'Open Data'. **'Open Data'** is defined by the Open Data Institute as *'data that's available to everyone to access, use and share'*. One of the key benefits of being a researcher using secondary quantitative data analysis is we live in an era characterised by a range of related, albeit distinct, initiatives seeking to create accessibility to data. These initiatives started within the physical sciences as far back as the 1950s with the aim of allowing a greater number of people to access and work with science data, not just to validate it but also to create a greater mass of people working on specific science 'problems'. This move-ment accelerated in the 1990s with the advent of better and faster computers and the internet. The digital revolution of the early nineties has further advanced these initia-tives, specifically in relation to social science data. Governments and other organisa-tions have embraced 'Open Data' as a means to show their commitment to greater transparency and accountability against the backdrop of the increasing collection of people's personal data in the digital world and the tightening of legal controls on personal data. 'Open Data' is linked to other initiatives such as 'Open Access', which is a commitment by academic journals and organisations to provide their work to all without access controls. You may also come across 'Open Science', which is a commitment to open access of findings, data and publications to all. Such initiatives seem positive and they mostly are, but we should not be naive to assume that all data is 'open'; organisations choose how much, how 'cleaned'/edited and who can access it. Nevertheless, the good news is that such initiatives mean we can more easily access the data that we want to work with.

Box 2.5

Reflective Exercise: 'Access All Areas'?

Most governments, organisations and businesses are committed to 'Open data', this means that their data should be freely and widely available to be scrutinised by the public. Usually such data is published online; indeed, the rise of the digital has driven 'Open data'. It is usually a statutory requirement of government agencies to do so, linked to greater legal protection of an individual's data and right to privacy.

Go online and research the debates around 'Open Data'.

Do you think that 'Open Data' makes government more 'democratic' and accountable?

Are there government data, which is not 'open access' and do you agree that it should be 'non-open'?

Can you identify criticisms of 'open data' policies?

How do you evaluate 'open data' initiatives?

Show me the data!

So far, we have identified and evaluated the differences between types of data and data collection. If you choose to conduct primary data collection, you obviously need to go out and collect it before you can do any descriptive statistics; you can skip the next sections and move straight on to the discussion of data management at the end. This chapter will not be discussing the process of primary data collection; instead, if you want to know more you should read Volumes 1 (*Beginning Quantitative Research*) and 2 (*Survey Research and Sampling*) in this series. However, if you have decided to pursue a piece of secondary data analysis, then you need to get your hands on some secondary data and the rest of this section will tell you how to do this. It has never been easier to access large amounts of data due to 'Open Data' policies; relatedly due to advances in digital media, we can literally access data from the comfort of our own laptop or similar device. Indeed, if you simply insert 'access data on [insert specific topic]' into an online search engine, you will be confronted by a long list of potential sources. The first step with every research project is to start with a general topic and basic research question that you are interested in and follow this with a review of the existing or most recent literature on that topic. Literature searches not only assist you in refining your research topic and ultimately formulating your research question but also shape your methodology and methods through review of previously conducted work. Volume 1 in this series, *Beginning Quantitative Research*, outlines how to conduct a literature review. By reviewing others' approaches to your topic, you will inevitably come across potential data sources that others have used and you should certainly have a look at them to see if they are accessible and useful to you.

A little exploration first

As well as being guided by others' work, another useful thing to do is to review data websites that have been designed for the public to access. The majority of Western nations have some form of state-funded statistical or data service, such as the ONS in the UK or the Statistisches Bundesamt (Federal Statistical Office) in Germany. Because of the adoption of 'Open Data' and 'Open Government' as a means for governments to appear more accountable, increasingly national data service websites are designed to be user-friendly with the non-data literate in mind. The UK's ONS (www.ons.gov.uk) is a good example of a user-friendly website that has lots of data visualisations and 'headlines' to allow the user to navigate around a range of national trends and data sets. It also contains access to specific data sets and related reports/publications. Most Western nations who are committed to 'open data' have similar national statistics websites, which you can use to navigate towards accessing specific data sets. In the USA, it is Data.Gov (www.data.gov); in the EU it is Eurostat (ec.europa.eu/eurostat?); in Australia, it is the Australian Bureau of Statistics (www.abs.gov.au); and in Canada, it is Statistics Canada (https://www.statcan.gc.ca). Transnational organisations also have embraced 'open data' and can be used to navigate data trends, including the United Nations (unstats.un.org/home/), the Organisation for Economic Co-operation and Development (data.oecd.org), the World Bank (data.worldbank.org) and the International Monetary Fund (www.imf.org/en/Data). However, we should note that not every 'open data' website is equally user-friendly. Some are more user-friendly than are others; compare the website of the UK's ONS with the International Monetary Fund's website and you will see what we mean. As a beginner to searching for data sets, you may have to do a bit of research into your topic before you start navigating a specific website.

In addition, you can explore specific data sets via online visualisation tools; for example, the UK Police (https://www.police.uk/) have a fantastic interactive crime map tool on their website which allows you to explore crimes and neighbourhoods, right down to street level; UK Census data can be explored visually on the Datashine (http://datashine.org.uk) website, which allows the user to explore Census data in specific locations. Another example is the UK's Ministry for Housing, Communities and Local Government funded 'Index of Multiple Deprivation' explorer (http://dclgapps.communities.gov.uk/imd/idmap.html), which allows you to explore levels of deprivation by specific locations. Chapter 6 in this book discusses data visualisation in more detail, but suffice to say, you could use such visualisation tools to get ideas about potential research questions and/or narratives and then you can go and search for the data itself, as each is built on freely accessible data. It may seem that exploring 'what's available' takes up precious time, especially if you are working to a deadline, but it is a useful stage because it allows you to see different statistical stories, trends and possible data sources.

Data, data, everywhere

As has already been highlighted, there are potential sources of data available from a myriad of organisations and locations. This can be overwhelming particularly to the novice researcher. Remember to let your research question and literature search guide you, and this will make your data search more targeted. It is also good to consider what sorts of data you are seeking and this will guide your search.

Research data falls into two categories: firstly, data funded by national governments and agencies in order to gauge changing public attitudes and behaviours, typically large-scale surveys (e.g. the GSS in the USA), and secondly, data deposited in data archives by researchers as part of the legacy of a research project. The latter may have more access controls on it than the former. Administrative data can be found on a range of sites from government funded to charities and commercial organisations, although the most commercially sensitive data will not be freely accessible. The type of data you choose to work with can influence the sort of analysis that you do, so it is important to be mindful of your research question and how the data is there to help you address it. If you read an interesting media report, citing statistics that interest you, then look for the reference to the original study as it could lead you to a potentially interesting data set. Using an online search engine can help identify potential sources of data as long as you are specific in your search phrases.

Accessing and downloading data

This book does not discuss how to choose a topic or formulate a research question, if you need help with that go and read Volume 1 in this series, *Beginning Quantitative Research*. This chapter presumes that you have a topic and a research question. Once you have decided on your topic and have narrowed down potential data sets, then you are ready to look at the available data. This means first going to the website where your data is stored; in this age of 'open data', you should not be asked for a fee to register or download data from a national data service. If you come across a demand for a fee, check with your course tutor as it may be an online scam. Most sites require that you register as a user and some give automatic access to students and academics, whereas non-academic users must apply directly to the data service for access. To illustrate, the UK Data Service (UKDS) provides automatic access to all students in higher or further education in the UK, meaning you access it through your institution and then complete a short registration form online. It is useful to be aware of different access protocols for different types of data. If data is categorised as 'Open' or 'Open Access' then anyone can download and use it. Sometimes the data won't have this label, it will just be downloadable. If the data is categorised as 'safeguarded' then that usually means the owner of the data wants it to be used for specific reasons, usually academic, non-commercial research. If your chosen data is labelled in this way, it just means you have to go through a couple of checks to explain how you will use the data. Typically, it will ask you to say something about your research project. Don't be put off if you see this on a data site; for example, the UKDS has a lot of data that is labelled 'safeguarded' but they just want you to provide a rationale for your usage. Your tutor can always help you in completing such forms. The most controlled form of data is categorised as 'restricted', 'sensitive', or 'controlled'. This may include data about child abuse, sensitive health data and/or data where the participants' identities may be at risk of disclosure due to small or highly localised samples. Whatever the ethical reasons for restricting the data, the data owners do not want 'open access'. 'Restricted' data is not data that is suitable for an undergraduate researcher; to access such data usually entails undergoing specific training as determined by the data owner and/or data service, which can take time. In addition, such data needs a greater awareness and training in ethics and research methodology than you would have at an undergraduate level. To access national data service–held data, you usually need to complete a registration form which often asks for your reasons for downloading the data; this is because they want to measure downloads and usage, so don't be put off by this. Once you've registered, your online account will store searches and data. Many sites will not require you to register as a user to download data, but if there is a registration option then take it because this will speed up your work. To illustrate, the UK's NOMIS website (https://www.nomisweb.co.uk/) is home to the UK's Census data (among lots of other data) and is hard to navigate, but if you register as a user, it remembers all your searches and therefore speeds up your data accessing.

Are you sure about this data?

Once you've found the data set that you think is for you, it is easy to rush to download. Before you do this, make sure that you have a really good look at the information held online about the data set. Many large surveys (see e.g. the BSA at https://bsa.natcen.ac.uk or the Crime Survey of England and Wales [CSEW] at https://www.crimesurvey.co.uk) have their own websites, which provide detailed information about the survey's history, variables asked, methodology and related studies. Some surveys' actual websites (see e.g. the US's GSS at http://gss.norc.org) host their data so you can review key information prior to downloading data.

Relatedly, when you download data, many websites give you the option to download accompanying documents, such as methodological information, variable lists and so forth. Read these documents first so that you understand the type of data and why it was collected. You don't want to download a large survey with 700 variables only to find that it doesn't actually contain the variables that you want. You would be amazed by how much students presume is in a specific survey; don't presume, check! In addition, reflect on why and how the data was collected. If it is a cross-sectional survey, do you have the most up-to-date version? If it is longitudinal, do you understand how to analyse it because such data can be challenging for a novice. If it is administrative, do you know by whom and why it was collected? Your tutor probably wants you to discuss some of these issues in the methodology section of your assessment, so use this information to help you. Also consider how others have used this data set; did they find it useful? Is the data available in a format that you can work with; many administrative data are in MS Excel or CSV formats. Once you are absolutely certain that this is the data set for you, then you can download it; some sites give you a choice of formats others do not, so be ready to possibly transfer data between formats.

Should I go large?

One last issue to consider prior to downloading some data is location and how large should you go? Because there is so much data available now, you can choose to be very specific with your focus. To illustrate, with UK Census data (but this applies to most censuses) you could conduct an analysis at the level of a neighbourhood (perhaps comparing an affluent district with a poor district within a city in relation to educational attainment) town/city, region or nation. Think of the different and compelling stories you could tell with such data. You might 'just' want to stick with looking at national data, perhaps views on gun control in the USA or Brexit in the UK. National data is possibly the easiest sort of data to work with because we feel most comfortable with it as the context is both familiar and interesting. You may want to compare trends across nations or even international blocs; for example, you could use the European Social Survey to compare work–life balance across all 27 EU states, which would be a big analysis! Alternatively, you might be interested in comparing work–life balance across specific regions of the EU, such as Scandinavia with the UK. You may even want to compare across very different nations; for example, you could use United Nations data to compare child health indicators across a range of nations or continents. As exciting as this sounds, you need to really think about the story that you want to tell with the data and the research question that you want to answer. Chapter 7 will help guide you with this. Just be mindful that the bigger the data set, the more variables and therefore the greater amount of data cleaning that you will have to do.

Downloading data

So we've said enough about reviewing sources of data let's actually look at downloading some data. This section is going to show you how to download data, using two examples. The first example is how to download data from the GSS and the second from the National Survey of Sexual Attitudes and Lifestyles (Natsal).

The general social survey

Beginning in 1972, the GSS is an American cross-sectional survey that seeks to monitor social change in the USA through collecting data on attitudes and behaviours. Since 1994, the survey is conducted every 2 years on even-numbered years. It is nationally funded, and after the US Census, it is the most used social science data in the USA. To access the GSS go to its website http://gss.norc.org/. The website provides information about the survey, updates on data releases and so forth. It also hosts the Quality of Working Life survey which is a national survey looking at Americans and work, which you may want to check out. There are two ways to access the GSS, and which one you choose depends on what type of analysis you want to work with.

Easy access to the gss without registering

If you just want to download the whole survey and you know which year that you want, then you can do this easily without having to register as a user (Figure 2.1).

Figure 2.1 Accessing the 'Get the Data' tab on the GSS website

Figure 2.2 shows the 'Get the Data' screen.

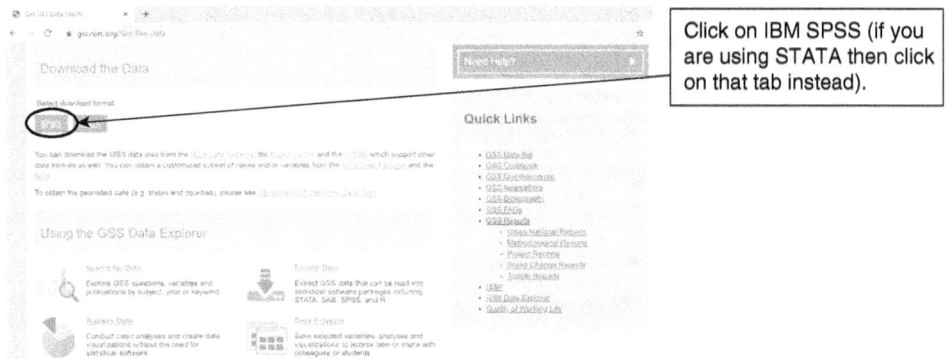

Figure 2.2   The 'Get the Data' screen on the GSS website

Clicking on the IBM SPSS tab takes you to the screen shown in Figure 2.3.

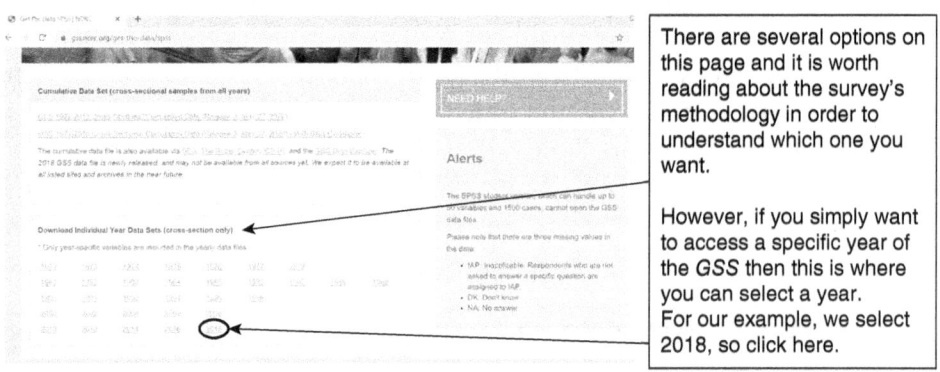

Figure 2.3   Selecting data on the GSS website

When you click on '*2018*', it prompts your computer to save a zipped folder (*2018_spss*); save this in a location of your choice. Then once it is saved, you can unzip the folder and you have your data.

Using GSS explorer to access data

If you choose to register for the GSS, then it allows you to make use of their fantastic 'GSS Explorer' site. You don't need to be a registered user to use the GSS Explorer, but it is a good idea as it will remember your analyses and data selections. First, you need to access the GSS Explorer, which is straightforward; start by going to the GSS website (Figure 2.4).

To go to GSS Explorer, click here.

Figure 2.4 Selecting GSS Explorer on the GSS website

Figure 2.5 shows the GSS Explorer website. It is worth you starting by creating an account, which involves completing a brief online form and literally takes a few minutes. To do this, click on the 'Create An Account' tab.

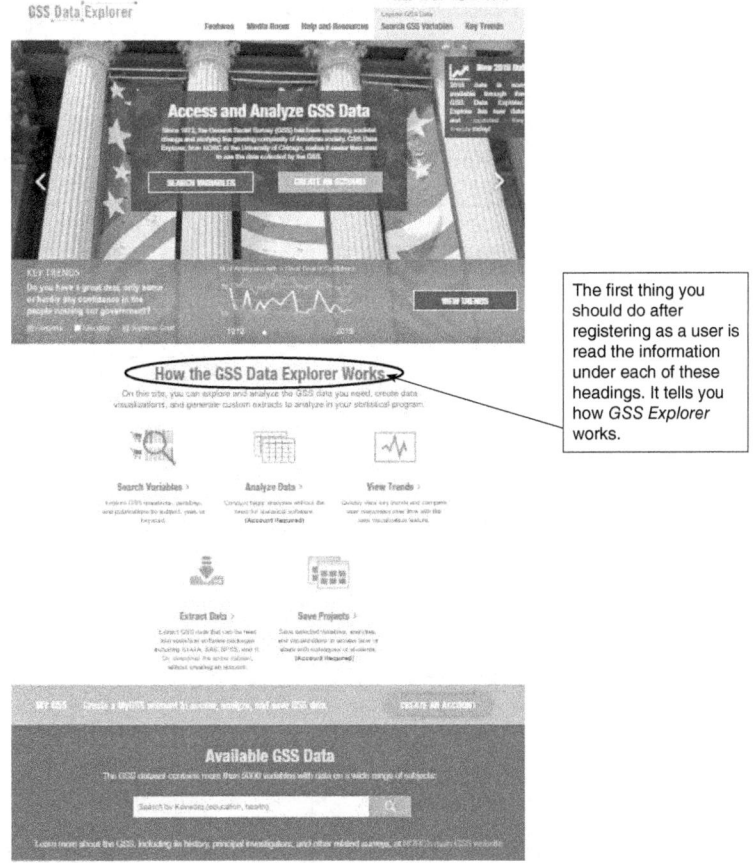

The first thing you should do after registering as a user is read the information under each of these headings. It tells you how *GSS Explorer* works.

Figure 2.5 GSS Explorer on the GSS website

Once you are confident that you know how GSS works, then you can plunge straight into accessing data. Alternatively, it has a great 'View Trends' button that takes you to a screen where you can play about with different variables and explore trends (Figure 2.6). You might want to do this as a means to familiarise yourself with the data, and it may aid you in formulating a research question.

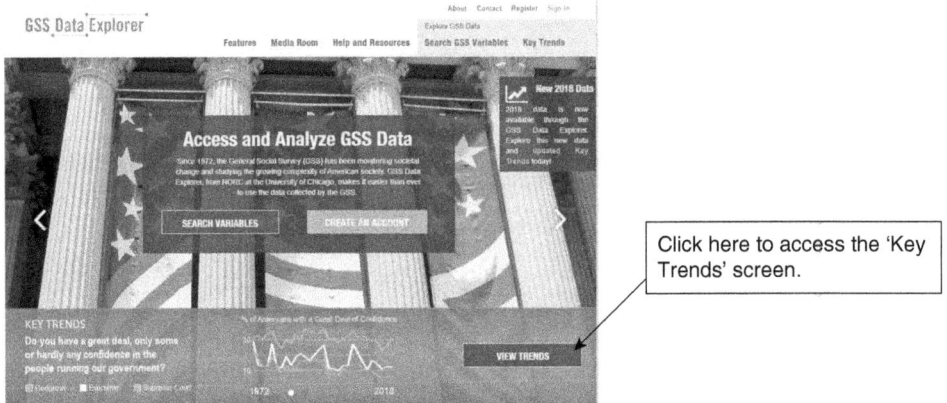

Figure 2.6   'View Trends' button on GSS Explorer on the GSS website

Once you click the 'Key Trends' button, it allows you to select a variable to explore and you can choose how you want to explore it. Figure 2.7 shows 'Is it wrong for same-sex adults to have sexual relations' by Age category.

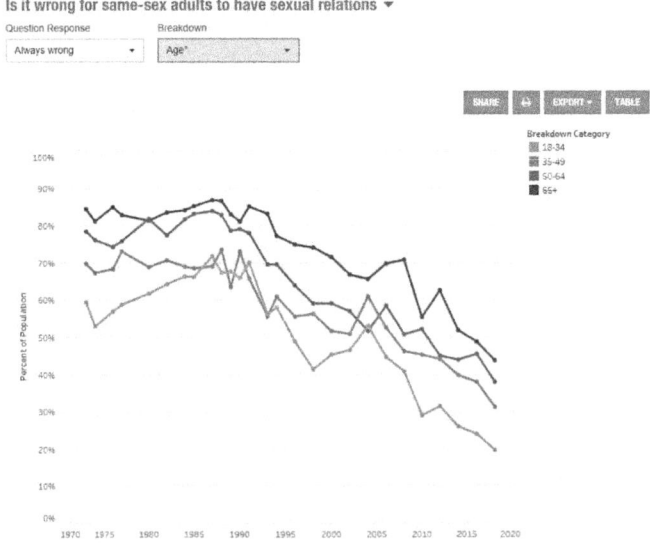

Figure 2.7   'Is it wrong for same-sex adults to have sexual relations' by age, GSS Explorer

This graph shows the trends since 1972. GSS Explorer allows you to print or explore this data. You can also show it as a table. You can export the graph as a PDF or the data on MS Excel. If you want to go straight to selecting variables, then click the 'Search Variables' button instead as shown in Figure 2.8.

Figure 2.8 Accessing variables on GSS Explorer

Figure 2.9 shows the search screen and you can choose to scroll through every variable (which is a bad idea) or you can search for the variables that you want, either using key words or topics. You can also filter by year. If you know the actual variable names (from the GSS codebook, also on the website), it will save time; you simply insert them into the 'Keyword' search.

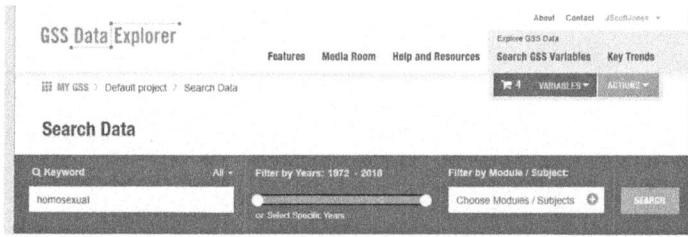

Figure 2.9 Using keyword search on GSS Explorer

In Figure 2.9, we insert the phrase 'homosexual' and click the 'search button'.

Figure 2.10 lists all the variables that feature the keyword. You then click the 'add to cart' button for the variables that you want. For our example, we will click '*homosex*'. Once clicked, it shows a tick, indicating it has been added to your cart. If you click on the variable name, it takes you to a frequency table for the variable and it shows other variables related to it that might interest you. Once you are happy with the variable(s) that you have chosen you can now do two things as Figure 2.11 shows.

The 'Analyze data' button allows you to analyse the data using GSS Explorer, which is great if you don't have access to statistical software. You can save, print and extract your analysis. Alternatively, if you want to download the data, then click the 'Extract data' button and it will guide you through the different steps to downloading your data:

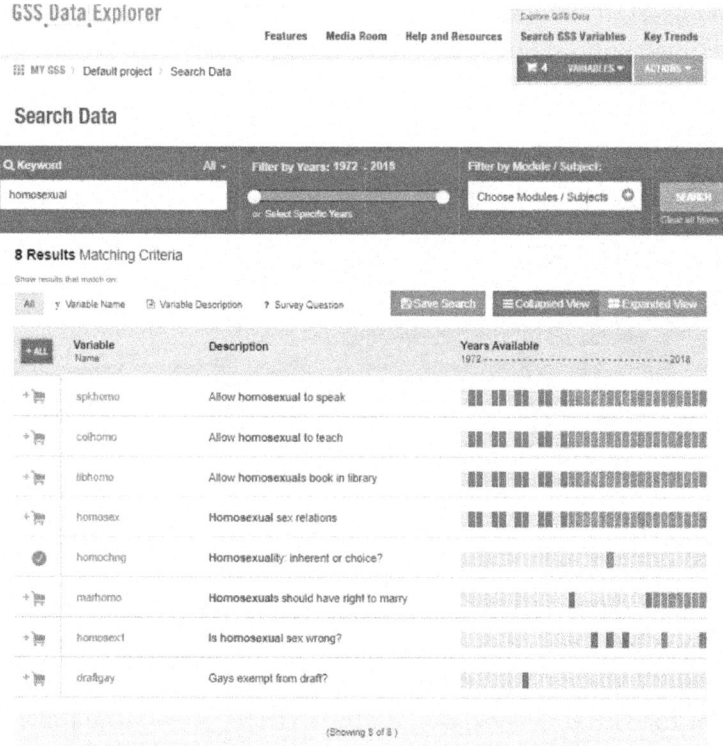

Figure 2.10   Results of keyword search on GSS Explorer

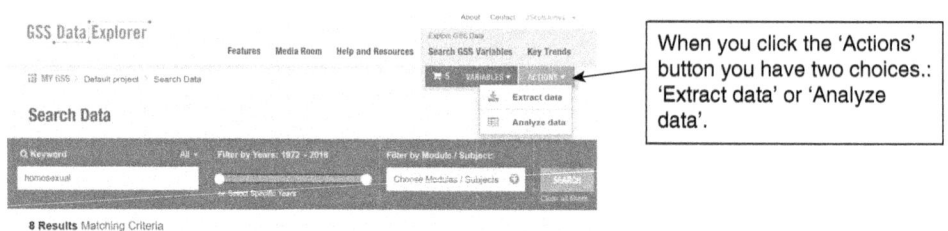

Figure 2.11   The Actions button on GSS Explorer

You can choose which variables you want and in what software format. Then you just go ahead and save the file to your computer.

The national survey of sexual attitudes and lifestyles 2010–2012 (natsal-3)

The GSS approach to building your data set through variable selection is commonly found on lots of other data service sites. However, some data service sites take a different approach, which we will illustrate using the Natsal-3 data set.

Natsal (http://www.natsal.ac.uk) is one of the largest surveys of sexual behaviour in the world and first started in 1990. It is conducted every 10 years (although it takes 2 years to conduct and complete), with the last one (Natsal-3) published in 2012. Natsal-3 is hosted by the UKDS (https://www.ukdataservice.ac.uk). The UKDS is one of the largest and best nationally-funded data archives in the world. It hosts the majority of the UK's major population surveys, including the CSEW and the BSA. The majority of data held by the UKDS is freely accessible to registered users. It is a well-designed site and even has a special section for student researchers (https://www.ukdataservice.ac.uk/use-data/student-resources) with lots of useful information about secondary data analysis; you can explore some of the data via an 'Explore online' button that leads to several online tools, prior to downloading it. There are few other data sites with the UKDS's level of student friendliness. You do have to register as a user in order to get the maximum functionality out of the UKDS. Figure 2.12 shows the front page of the UKDS website.

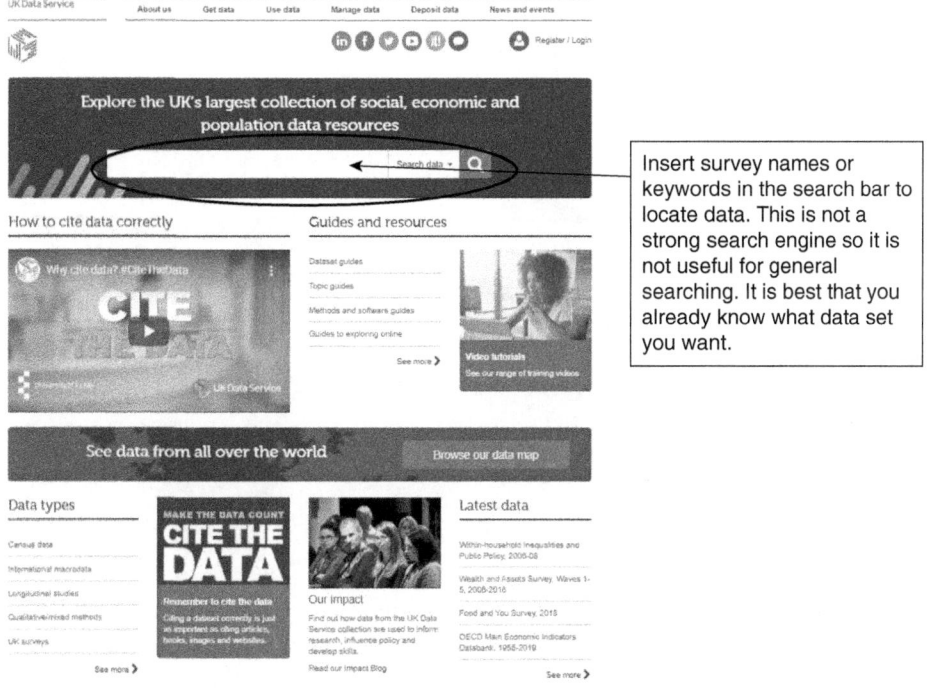

Figure 2.12 The UK Data Service website

If you insert Natsal 2010–2012 into the search engine, then it will take you to a list of all the Natsal surveys, you then click on the one that you want and it takes you to the screen shown in Figure 2.13.

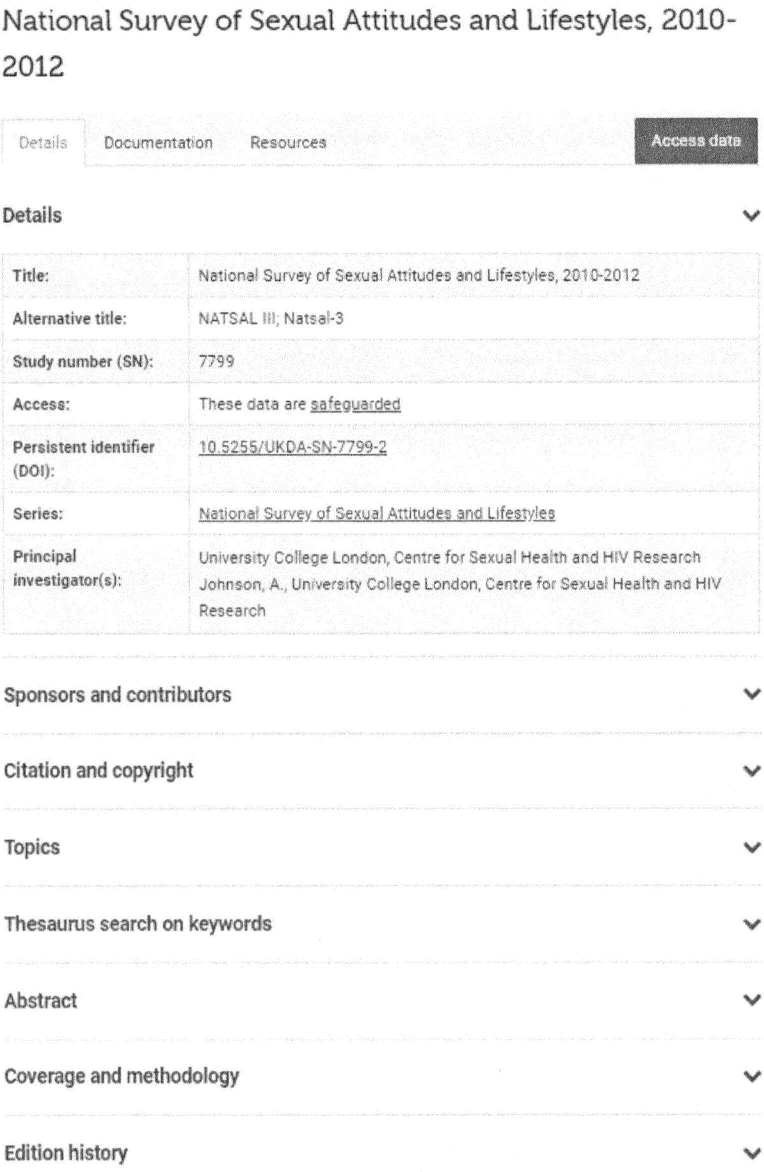

National Survey of Sexual Attitudes and Lifestyles, 2010-2012

| Details | Documentation | Resources | Access data |

Details ⌄

Title:	National Survey of Sexual Attitudes and Lifestyles, 2010-2012
Alternative title:	NATSAL III; Natsal-3
Study number (SN):	7799
Access:	These data are safeguarded
Persistent identifier (DOI):	10.5255/UKDA-SN-7799-2
Series:	National Survey of Sexual Attitudes and Lifestyles
Principal investigator(s):	University College London, Centre for Sexual Health and HIV Research Johnson, A., University College London, Centre for Sexual Health and HIV Research

Sponsors and contributors ⌄

Citation and copyright ⌄

Topics ⌄

Thesaurus search on keywords ⌄

Abstract ⌄

Coverage and methodology ⌄

Edition history ⌄

Figure 2.13 The 'Details' tab for Natsal-3 on the UKDS

Note. Natsal = National Survey of Sexual Attitudes and Lifestyles; UKDS = UK Data Service.

There are three tabs available, the one shown is the 'Details' tab. This provides a wealth of information, including citation, topics, methodology and an abstract. Take the time to read through this information. When you move to the 'Documentation' tab you get a new screen as shown in Figure 2.14.

Details | Documentation | Resources | Access data

Documentation

Title ▲	File name	Size (MB) ⇕
National Survey of Sexual Attitudes and Lifestyles, 2010-2012 Web Data File	eul_level_natsal_2010_web_data_for_archive_ukda_data_dictionary.rtf	0.25
Natsal-3, 2010-12: Codebook	7799_natsal_3_codebook.pdf	2.01
Natsal-3, 2010-12: Note for Users	7799_natsal_3_user_note.pdf	0.42
Natsal-3, 2010-12: Note for Users of Web Data	7799_natsal-3_web_data_user_note.pdf	0.22
Natsal-3, 2010-12: Questionnaire	7799_natsal_3_questionnaire.pdf	0.56
Natsal-3, 2010-12: Technical Report Volume 1: Methodology	7799_natsal_3_technical_report_vol_1.pdf	1.61
Natsal-3, 2010-12: Technical Report Volume 2: Appendices	7799_natsal_3_technical_report_vol_2_appendices.pdf	3.17
Natsal-3, 2010-12: Web Comparison Questionnaire	7799_natsal-3_web_questionnaire_annotated_for_data_users.pdf	0.73
Study information and citation	UKDA_Study_7799_Information.htm	0
UK Data Archive Data Dictionary	eul_natsal_2010_for_archive_ukda_data_dictionary.rtf	0.9
UK Data Archive Information for Study 7799	read7799.htm	0

Figure 2.14   The 'Documentation' tab for Natsal-3 on the UKDS

Note. Natsal = National Survey of Sexual Attitudes and Lifestyles; UKDS = UK Data Service.

This screen features links to PDFs. Each PDF provides important information about the survey, including the Codebook, questionnaire and technical report. When you move to the 'Resources' tab you get a new screen as shown in Figure 2.15.

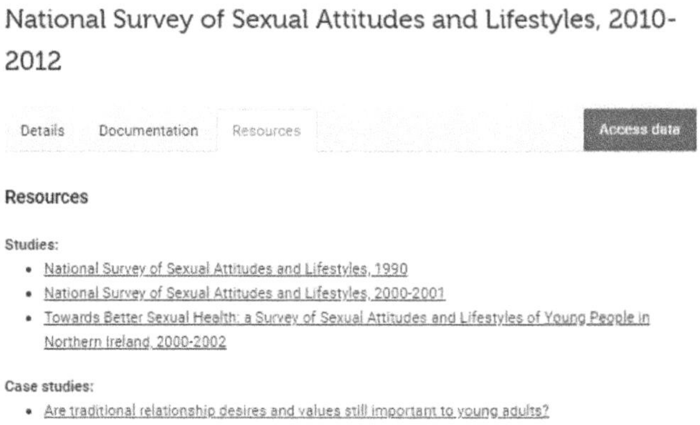

National Survey of Sexual Attitudes and Lifestyles, 2010-2012

Details   Documentation   Resources   Access data

Resources

Studies:
- National Survey of Sexual Attitudes and Lifestyles, 1990
- National Survey of Sexual Attitudes and Lifestyles, 2000-2001
- Towards Better Sexual Health: a Survey of Sexual Attitudes and Lifestyles of Young People in Northern Ireland, 2000-2002

Case studies:
- Are traditional relationship desires and values still important to young adults?

Figure 2.15   The 'Resource' tab for Natsal-3 on the UKDS

Note. Natsal = National Survey of Sexual Attitudes and Lifestyles; UKDS = UK Data Service.

This screen features links to other versions of the Natsal survey that you might be interested in. It also links to studies that have used Natsal data, which you might want to read and include in the literature section of your assessment. Take the time to read all the materials on the website before you hit the 'Access data' button. When you do hit the 'Access data' button, you are taken to the screen shown in Figure 2.16.

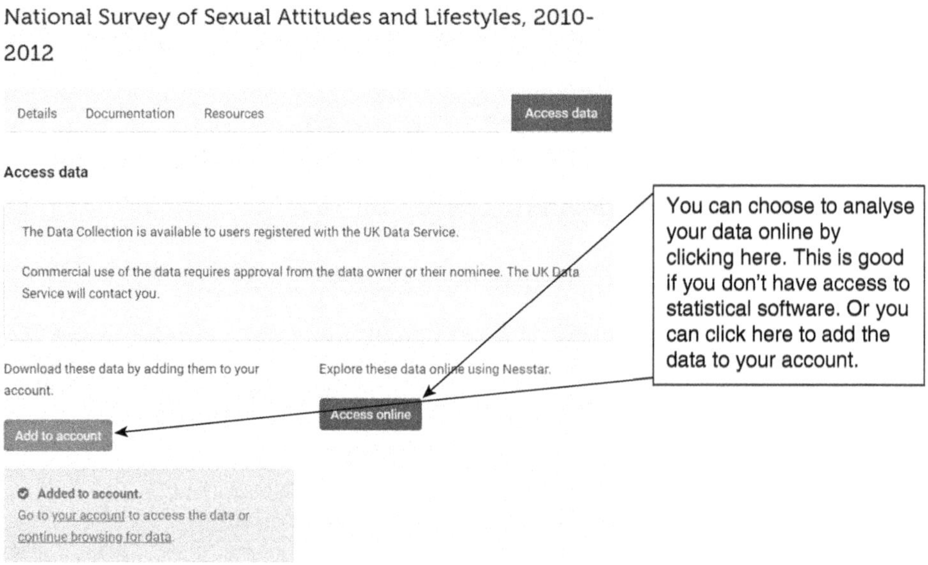

National Survey of Sexual Attitudes and Lifestyles, 2010-2012

Details   Documentation   Resources   Access data

Access data

The Data Collection is available to users registered with the UK Data Service.

Commercial use of the data requires approval from the data owner or their nominee. The UK Data Service will contact you.

Download these data by adding them to your account.   Explore these data online using Nesstar.

Add to account   Access online

You can choose to analyse your data online by clicking here. This is good if you don't have access to statistical software. Or you can click here to add the data to your account.

✓ **Added to account.**
Go to your account to access the data or continue browsing for data.

Figure 2.16   The 'Access data' button for Natsal-3 on the UKDS

Note. Natsal = National Survey of Sexual Attitudes and Lifestyles; UKDS = UK Data Service.

If you decide you want to download the Natsal-3 data, then you must add it to your online account. You do this by adding it to a 'project' that you have set up within your account; it could be 'Research report'. Figure 2.17 shows an example of a set of data sets attached to one project.

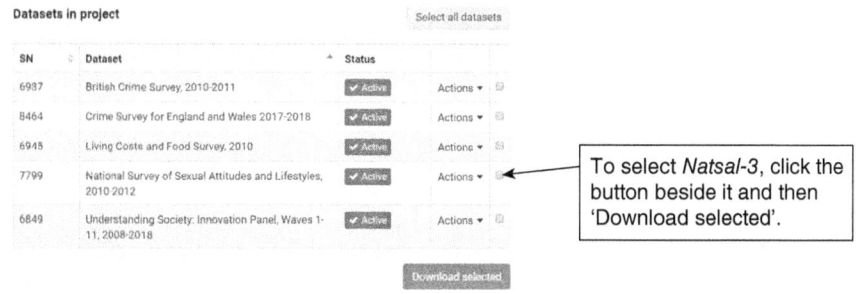

Figure 2.17 Data sets listed within a project on the UKDS

Note. UKDS = UK Data Service.

Figure 2.18 shows the download screen where you can choose a software format. You simply tick the format that you want and hit 'Download selected'.

Download

Unselect all downloads

File Format	File Size (mb)	Download	
Dataset: National Survey of Sexual Attitudes and Lifestyles, 2010-2012			
SPSS	11.2	Download	☑
STATA	11.14	Download	☐
TAB	11.97	Download	☐

Download selected

Figure 2.18 The Download screen on the UKDS

Note. UKDS = UK Data Service.

Once you hit 'download' it prompts you to save a zipped folder, go ahead and save the folder and then extract all. Figure 2.19 shows the inside of the folder.

Name	Type	Size
mrdoc	File folder	
spss	File folder	
7799_file_information	Rich Text Format	9 KB
read7799	Chrome HTML Docu...	3 KB

Figure 2.19 The downloaded Natsal-3 folders

Note. Natsal = National Survey of Sexual Attitudes and Lifestyles.

There are two folders. The '*mrdoc*' contains all those PDF files from the website, so keep these safe for future reference. Don't print them as some are very large. The '*spss*' folder is the actual data, which you can now start analysing.

Very few data services provide the level of detail that the UKDS does, most are similar to the GSS. The UKDS does provide an astonishing array of information alongside its data, so it is certainly worth exploring. They also provide a handy online citation tool for all data sets on each data set's catalogue page (making referencing them easy).

Box 2.6

Time to Get Your Hands Dirty! Finding and Downloading Data Sets

Now that we've shown you two different sites from which you can download data, it is time for you to have a go. Why not go to the GSS website and either download an entire data set or else take on the challenge of identifying a couple of variables and downloading them instead. Alternatively, you could fly solo and have a go at downloading some data from the data service of your choice.

Got data: let's look inside!

We are going to use the Natsal-3 data set, obviously you may be using a different data set, but the basic principles and issues that we will discuss in this and subsequent sections will be true of any data set. We should also note that we use IBM SPSS in our examples, unless otherwise stated, so if you are using a different software then your outputs or screen will look different. We are not allowed to show screenshots of '**raw data**' from Natsal-3 in this book, but we can describe what you would encounter if you opened it in IBM SPSS. In 'Variable View', you would see 1224 variables (!) and 15,162 cases or respondents (!) if you scrolled down in 'Data View'. It is big!

It is normal to feel overwhelmed when you have first sight of these data sets, particularly as you are probably used to nice and tidy small data sets that your tutor(s) have created for you. To help you with getting control of a data set, here are some tips:

- Read everything you can about the survey, its methodology and variables; make sure you review the codebook.
- Customise (if you can) the software 'view screen'.
- Move (if you can) the variables that you have chosen to the top of any 'view screen'.
- Delete all administrative variables as you do not need these for analysis, for example, 'scrambled serial number' or 'year of interview'.
- Delete all variables that you do not need for your analysis.
- Topics and/or themed questions are usually grouped together.

- Recoded or regrouped variables are often at the end of large data sets.
- Demographic variables (sex, age, ethnicity) are usually interspersed throughout a data set.

Problem buster!

There are many problems that you may encounter with working with secondary data, especially at the early preparation and cleaning stages; here are a few of the common ones:

- *I don't want to delete the wrong variables!* The majority of software packages have an 'undo' button if you make a mistake. Alternatively, if you follow the 'Versioning' approach outlined in the next section, then you will be able to go back a step and retrieve that variable.
- *I don't know which version of a variable to choose?* In large data sets, you will find that some variables have several versions. For example, if you look in Natsal-3, there are four different variables relating to age (*dage, dagegr, agrp* and *agrp2*), which one you select depends on the type of analysis that you want to do. By running descriptive statistics, you can explore each one and find the one most suitable for your project.
- *I can't decide which variable!* Although you should have a working list of variables prior to looking at a data set, you may, as you look inside it, find yourself drawn to new variables. The key thing is to consider which variable is best for your research question, run descriptive statistics and review key literature, as these will guide you. You may also need to consider what sort of tests you want to run, as this will shape the type of variables selected.
- *I don't have the right file format!* The large data services sites typically offer data in a range of software formats. However, you may find that the data you want is in a software with which you are not familiar. Fortunately, the majority of statistics software has the capability to transfer between software formats; thus, MS Excel can be converted to IBM SPSS (and vice versa), R can accept IBM SPSS and .csv and so forth. The key thing is to ensure that the data is in the correct condition to transfer – for example, if you are transferring MS Excel to IBM SPSS, then the variables have to be in vertical columns and the cases in horizontal rows. There are loads of videos online that talk you through data transfers between software formats.

Shit in, shit out: and other key principles of data management

Now that we've looked inside our data set, we now need to think about data management. **Data management** is a catch-all term that refers to all the elements that go into a research project, including ethics, storage, cleaning and analysis. When you do primary data collection, then data management starts at the very beginning of the project and shapes it. When you are dealing with secondary data, you don't need to think about data management until you literally have downloaded your data. There are a couple of key principles to remember in relation to data management and you would do well to remember them:

- *Shit in shit out.* Also known as 'garbage in and garbage out', this rule reminds us that if we don't take the time to clean and prepare our data, then no matter how good our analysis may be, the end result will be poor. In any research project using secondary data, the two longest stages should be literature survey and data preparation.
- *Trust no one.* You may assume that data that you access from a data service is clean and error free – you would be wrong. Most data needs some element of cleaning not just because it will have errors but because you may want the data to be 'cleaned' in a specific way to aid your analysis.
- *Murphy's Law.* Murphy's Law dictates that what can go wrong will go wrong and usually when you least want it to. A classic example of this is the campus network crashing just as you are about to submit your assessment, 3 minutes before the deadline. If only you had submitted it earlier then you would have avoided this mishap. In terms of data management, a good way to mitigate against Murphy's Law is to save and back up your work and make copies of these backups. Additionally, you should adopt 'Versioning', which means saving a version of the data every time you 'clean' or amend it. Thus, the first downloaded data, which is uncleaned, could be labelled 'Unclean version' or 'raw' version'; the first time you delete some variables could be 'Version 1 – variable remove'. What you call each saved file doesn't matter as long as you know which version it is; if you make a mistake such as deleting the wrong variable 'versioning' allows you to go back a step and start again. If you don't use 'Versioning' then when you make an error you have to return to the original data, wasting time and effort. You may want to make a data log to itemise each version as you go along to keep track of what you have done.

Some basic cleaning tips

Data cleaning involves many processes and probably deserves its own book. Nevertheless, there are some basic things you should always do. The first is correct the labelling of your data; often the 'Measure' is incorrectly labelled (e.g. IBM SPSS has a tendency to default everything to 'Scale') and the codes for the 'Missing' categories have not been applied (as is the case for our Natsal-3 data). These sorts of issues can be easily resolved and take little time. Second, when you run frequency descriptives for each variable, then you will soon see any errors for categorical data; you can double-check this with the codebook too and then simply clean the data accordingly. With Scale data, there is the challenge of identifying which are genuine errors and which are outliers within the data. Let's illustrate this with an example from the Natsal-3 data set. The variable we will use is *periods2*, which asks 'Age when started menstruating'. Figure 2.20 shows the variable in IBM SPSS 'Variable View', please note that we have moved the variable to the top of the data set (along with *dage*) for this example.

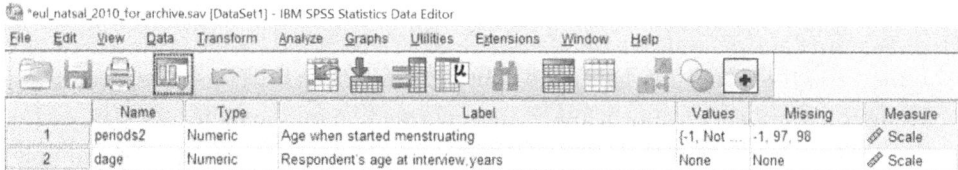

*eul_natsal_2010_for_archive.sav [DataSet1] - IBM SPSS Statistics Data Editor

File Edit View Data Transform Analyze Graphs Utilities Extensions Window Help

	Name	Type	Label	Values	Missing	Measure
1	periods2	Numeric	Age when started menstruating	{-1, Not	None	Nominal
2	dage	Numeric	Respondent's age at interview,years	None	None	Scale
3	sin2	Numeric	Scrambled serial number	None	None	Scale

Figure 2.20   periods2, Variable View, IBM SPSS

You should instantly spot that the variable has been incorrectly labelled as 'nominal' and that no 'Missing' values have been inserted. So let's go ahead and change that, as Figure 2.21 now shows.

*eul_natsal_2010_for_archive.sav [DataSet1] - IBM SPSS Statistics Data Editor

File Edit View Data Transform Analyze Graphs Utilities Extensions Window Help

	Name	Type	Label	Values	Missing	Measure
1	periods2	Numeric	Age when started menstruating	{-1, Not ...	-1, 97, 98	Scale
2	dage	Numeric	Respondent's age at interview,years	None	None	Scale

Figure 2.21   'cleaned' periods2, Variable View, IBM SPSS

If we run MCT for *periods2*, we can have a look at the data more clearly. Table 2.1 shows the MCT for our variable.

Table 2.1   MCT, periods2, IBM SPSS

Statistics

Age when started menstruating

N	Valid	8774
	Missing	6388
Mean		12.89
Median		13.00
Mode		13
Std. Deviation		1.581
Variance		2.501
Range		19
Minimum		7
Maximum		26
Percentiles	25	12.00
	50	13.00
	75	14.00

Note. MCT = measures of central tendency.

The mean, median and mode are all very close together, the standard deviation is low and the variance is low: this all suggests that the majority of the scores lie between Age 12 and 14.5 years, with little variance in the data. There are much lower and higher scores, indeed the range is 19, which is quite high, we would need to explore this data further to decide whether these low or high scores were outliers or part of the distribution. One technique is to apply the *Outlier Labelling Rule*, whereby you use the following equation to identify which scores fall outside of the normal distribution and are therefore outliers:

$$SUM = (Q3 - Q1) * g \ [g = 1.5]$$

For our data this would be:

$$(14 - 12) * 1.5 = 3$$

You then add the 3 to Q3 (14 + 3) and subtract the 3 from Q1 (12 −3), which means that anything above 17 and below 9 are identified as outliers and not part of the normal distribution. You then have to decide what to do about this, for example, by removing these scores through making them 'missing', which is known as 'data trimming' or you could Winsorize them by changing all the scores below 9 to 9 and all the scores above 17 to 17, which puts them within the normal distribution. Tables 2.2 and Table 2.3 show the MCT for our 'cleaned' data.

Table 2.2 MCT for winsorized periods2

Statistics

Age when started menstruating

N	Valid	8774
	Missing	6388
Mean		12.89
Median		13.00
Mode		13
Std. Deviation		1.554
Variance		2.416
Range		8
Minimum		9
Maximum		17
Percentiles	25	12.00
	50	13.00
	75	14.00

Note. MCT = measures of central tendency.

Table 2.3 MCT for 'trimmed' periods2

Statistics

Age when started menstruating

N	Valid	8749
	Missing	6413
Mean		12.88
Median		13.00
Mode		13
Std. Deviation		1.541
Variance		2.374
Range		8
Minimum		9
Maximum		17
Percentiles	25	12.00
	50	13.00
	75	14.00

Note. MCT = measures of central tendency.

We can see from Tables 2.2 and 2.3 that the MCT are not radically changed, despite our cleaning methods, but what we now have is approximately normal data. These techniques are controversial, but you could say that all manipulation of our data changes it to some degree and is therefore controversial. The important thing is that we always make it very clear when we write-up our analysis what data cleaning we have done and why. You will find that you gain extra marks in an assessment when you clearly discuss your data cleaning rational.

Looking ahead

In order to fully clean and prepare data for analysis, whether at the descriptive or inferential level, we need to know and understand the different types of variables and therefore data that we may come across. If you know you're categorical from your interval-level data, then you can go ahead and skip to later chapters of this book. However, if you are less sure of the differences between types of data and what that might mean for both cleaning and analysis, then the next three chapters are for you.

Chapter Summary

- The first decision we must make is whether we want to collect our own data (known as primary data collection) or use data collected by other researchers (known as secondary data collection).

(Continued)

- The advantage of research using primary data is that researchers shape and control the whole process including design and distribution; however, it can be costly and require a lot of time.
- Alternatively, working with secondary data is time- and cost-efficient. Nonetheless, it is someone else's data, collected for a specific purpose, without your research project in mind.
- We can categorise secondary data into two broad categories: research data and administrative data. Research data is data collected for the sole purpose of academic research and administrative data is data collected by a variety of organisations.
- Most governments, organisations and businesses are committed to 'Open data', this means that their data should be freely and widely available to be scrutinised by the public.
- When you download data, many websites give you the option to download accompanying documents, such as methodological information, variable lists and so forth. Read these documents first so that you understand the type of data and why it was collected.
- When you do primary data collection, data management starts at the very beginning of the project and shapes it.
- The important thing is that we always make it very clear when we write up our analysis what data cleaning we have done and why.

Further Reading

MacInnes, J. (2017). *An introduction to secondary data analysis with IBM SPSS statistics.* Sage.

If students want to take their statistical skills to the next level when using someone else's data, then take a look at this book. It deals with many more statistical techniques in a very approachable way with some great examples.

3

MEASURE EVERYTHING – LEARN SOMETHING – ANSWER NOTHING

AN EXPLORATION INTO VARIABLES AND TYPES OF MEASUREMENT

Chapter Overview

Introduction 56

Measurement as a taken for granted 57

Measuring the social world 60

Units of measurement: variables 64

Levels of measurement 72

Categorical variables... 74

Interval variables 74

Looking ahead.. 77

Further Reading 78

Introduction

In Chapter 2, we reviewed different sources of data that were freely available for analysis; additionally, we examined the differences between these data, linked to why and how they were collected. However, before we can get stuck in and start conducting analysis of our data, we need to consider what it is we actually want to find out. In Chapter 1, we discussed what descriptive statistics are and how useful they can be for us, not just in identifying trends and patterns in our data but also because they can help us clean and explore our data for more complex analysis. Relatedly, we noted the difference between descriptive and inferential statistics – the former describes the sample and the latter is used to test hypotheses and potentially make generalisations from the sample back to the population. Nevertheless, both approaches to statistical analysis require us to consider the issue of **measurement**: why, on average, are men paid more than women. This is both a measurement of 'pay' and a measurement of 'sex'. Think about educational underachievement amongst children. Did you know it varies by class and ethnicity in many nations, including the UK and the USA? Again, these are all measurements ('educational achievement', 'class' and 'ethnicity') just as 'height', 'weight' and 'time' are; although the latter seem more intuitive to us as they are familiar and we are used to them as measures, so they seem like they are actually measuring something tangible. However, we should note that how something like weight is measured varies through history and across cultures, but we probably accept that there are specific standard measures of weight that we are familiar with, such as kilograms and grams.

When we begin to think about the concept of measurement, it can be confusing in the social sciences, because we are not measuring something as observable or as easy to concretise as height, weight or time to the point that what we do measure will always have an element of contestability (more on this later in the chapter). Perhaps it is this contestability that makes the social sciences so exciting in that we are never truly sure of anything. The best we can do is find evidence that supports our thinking and evidence that does not. As Euripides, an ancient Greek philosopher, once said, '*Question everything, learn something and answer nothing*'. We would rather add 'measure everything' as it is this that is the bedrock of social science research: And just as we measure and learn more, it often reveals just how little we actually knew. But hey, life's complicated, what do we expect?

In this chapter, we will take you on a journey of measurement and hopefully reveal just how important this is when considering exploratory and descriptive statistics. Measurement is an important part of the social science process of classifying, counting, measuring and explaining. All four elements are important and will be explained in more detail later in the chapter. The central part of the adage has to be to '*learn something'*, well said Euripides! As such, this chapter will explore the issues surrounding the measurements that social scientists use in their attempt to explain the social world, along with the different types of variables used that can help categorise and then explain that which we have perhaps foolishly attempted to measure.

Measurement as a taken for granted

Initially, the concept of measurement may seem straightforward and so pervasive in our everyday life that we may rarely give much thought to the extent to which we quantify everything and anything. Contemporary modern society appears preoccupied with measuring stuff: whether it be the weight (measured in kilos or pounds if you are under the age of 40, or pounds and ounces if you're being traditional and probably well over 40) of a newborn baby, the time (measured in minutes and seconds or even nanoseconds) it takes Usain Bolt to run 100 metres or even the speed at which our mobile broadband works to the Mbps (measured as the speed at which data is transferred every second). Indeed, the second thing most people comment on at the birth of a baby (after identifying its sex) is its weight, with both medical ('overweight', 'normal' and 'underweight') and cultural ('healthy') evaluations made about the newborn based on this measurement. We measure so much that we might take for granted the complexity of the actual task, along with the vast array of different units of measurement we have at our disposal. Before we move on to discuss measurement in social research, let's look at how we measure something perhaps more familiar to us – distance.

There are a large number of ways open to us that can measure distance or the physical size of something (length/breadth/depth). You will most likely be familiar with the metric system that uses millimetres, centimetres, metres and kilometres. The metre is a universal system of measurement that comes from the Greek noun μέτρον, which means literally *'measure'*. This, however, is not the only way to measure the distance between two points. In the UK there is a mishmash of different measurements used depending on what distance we are measuring. For instance, if we wanted to know the distance between the cities of Manchester and London, you would instantly know not to use a ruler (a measurement tool commonly used in the classroom for short lengths) and you would most likely not use the metric system in the UK. Instead, you would use the Imperial measurement of miles (1 mile is equivalent to 1.609 kilometres) because the UK uses an odd mixture of the Imperial and metric systems. Because we know how many kilometres there are in a mile, it is easy to make the conversion between the two systems: the distance between Manchester and London is 208.5 miles or 335.5 kilometres (assuming we know which part of Manchester and London are the points at which we start and stop the measurement). We could also use the metric system in the first instance and dispense with the Imperial measure, which for most other things is the standard by which we measure in the UK. Relatedly, knowing what is being measured is also important as it then indicates the appropriate unit of measurement to use if not which system of measurement to use. Using the example of the distance between Manchester and London, it would be equally valid to use centimetres or even millimetres[1] (or yards and inches for that matter), but we know that for measuring long distances, it is not appropriate to use the smaller units. The point of measuring anything is so that we have information that is useful to us.

We might be forgiven for thinking that measuring distance is a precise endeavour, which it can be, but by no means has to be; indeed, we often describe quantifications using more concrete examples. For instance, how many times have football fields or double-decker buses been used as a measurement of size in the media? Look at Figure 3.1 from Liszewski (2015) who uses the size of football fields to illustrate the size of the world's largest container ship.

World's Largest Container Ship Has Four Football Fields of Deck Space

Figure 3.1 Container ship

As a 'rule of thumb' (which is itself the Victorian measure used to indicate the thickness of the stick with which a man was permitted to beat his wife – which should be no wider than the thumb of the husband), we do not use the size of a football pitch as a precise measure, not least as there are no standard sizes to a football pitch. Nevertheless, did you know that it would take 115,584 double-decker buses (which do have a standard size in the UK, Figure 3.2) laid end to end to reach from Land's End to John O'Groats (using the double-decker bus calculator – http://chrico.mazca.com/calc-lengths.html)

---

[1]Because we know you want to know, the distance between Manchester and London is 335,500000 millimetres.

Figure 3.2 Double-decker bus

Perhaps what is more surprising is that there is actually a double-decker bus calculator (thanks to Christopher Cockbill, you're a genius)!

So football fields and double-decker buses are broadly measuring the same thing – large objects or great distances – although it is doubtful whether you will ever hear someone's height being described as percentage of a football field or a double-decker bus:

> Oh hasn't he grown, he's nearly 1/5ths of the number 43 bus . . .

We could expect, however, a more standardised measure for height that offers a little more precision than the double-decker bus can offer. This could be a tape measure, a large ruler or even a laser measure if we want to give an air of even more accuracy to the nanometre (which is one thousand-millionth of a metre). Pretty accurate you might think, and you would be correct; but this is only because you know the metric system and what it can appropriately measure. Of course, a 30-cm ruler is the right thing to use to measure short distances between two objects or the length or width of something relatively small, such as a mobile phone. Yet it would be pretty rubbish at measuring its weight or temperature. That is because different measurement tools are used to measure different things. For weight, we can use a set of scales; for temperature a thermometer and so forth. These are all **valid** measures; that is, they are able to measure what they set out to measure in the same way that a ruler is an **invalid** measure of temperature. **Validity** is an important concept in social research and is linked to the concept of **reliability**, which relates to the effectiveness of a tool of measurement; a thermometer is a reliable tool for measuring temperature because different people could all use the same thermometer to measure a room's temperature and they would all get the same result. However, a thermometer would not be a valid measure of length. So far, we have focused on ways to measure the physical world; clearly, people have sought to quantify it for millennia. Such units and tools of measurement may seem intuitive and unproblematic, but we should be mindful that by adopting a historical perspective we can see that how we measure the physical world has changed many times and reflects cultural as much as physical contexts. Let's use days of the week as an example. Have you ever wondered why there are 7 days in a week? The length of the day (24 hours) is more understandable as this is the time it takes for a full rotation of the Earth on its axis. But 7 days is much less easy to explain. The route to the 7-day week has its origin in Roman, Babylonian and Jewish culture. The Romans initially had 8 days in a week with the market being held on the eighth day. The Babylonians had a 7-day week named after the then known planets (Sun, Moon, Mars, Mercury, Jupiter, Venus and Saturn). The Roman Empire and ancient Judaism adapted this 7-day a week calendar, which eventually was adopted by much of the world as we now know. I can almost hear you yawn at such a revelation, but wait for a few moments and remember that part of the job of a social science researcher or student is to question everything, especially the taken for granted such as a 7-day week. Once you do, you might see that we are actually stuck with a measurement legacy that, if questioned, doesn't actually make much sense. Let's begin with the number of days and weeks there are in a year: 365 days or 52 weeks. This is the time it takes Earth to go around the Sun (let's ignore the quarter extra day and leap year for the time being). Everybody knows this. But should we do the maths? If there are 7 days per week and 52 weeks per year:

$7 \times 52 = 364$ (gasp – let's check this again! Hmmm). Something does not seem quite right. It transpires that there are 7 days a week, 52 weeks per year and 1 day! If you include the quarter extra day added together for a Leap year, we then have 7 days, 52 weeks plus 2 days! We all knew this implicitly as this is why our birthday always falls on a different day each year.

We are the victim of measurement legacy that we take for granted and seldom question. Apparently, the Babylonians came up with the calendar to chart the movement of the moon which typically takes around 28 days to circle the earth. Interesting to note that 28×13 is 364, perhaps we should have 13 equal sized months plus one day? Who knows? The point is, the tools we have to measure anything, be it days, months or years, or kilometres, miles or buses, are often imprecise, but they are the tools we have to work with until better ones are found. The same imprecision applies to our feeble attempts to measure aspects of our social world.

Measuring the social world

As previously noted, there is a good deal of cultural understanding at play here for us to know what measures what. It does not get any easier once we start to measure things in the social sciences. As illustrated above, measurement has varying degrees of sophistication and levels of precision. With social research, however, it is more difficult to achieve precise measurements due to the complexity of what is being measured. This chapter explores measurement as a general topic, if you are interested in the specific issues relating to measurement in the design of questionnaires, then you should read Volume 2 in this series (*Survey Research and Sampling*), and if you want to know more about measurement in experiments, then go and read Volume 4 in this series (*Experimental Design*). As we noted in Chapter 2, the earliest forms of social data were administrative data collected by those in power for two central reasons: firstly, to literally stocktake what they had, whether people, land, gold and so on, usually so that taxes could be raised. Secondly, to determine levels of social support or services; for example, how many houses needed building or poor relief to be allocated. A good example of this is the *Domesday Book*, which was essentially a survey of England commissioned by William the Conqueror in 1086. What he was interested in was how much land and people were in England, so he knew how much tax could be raised. This may seem straightforward, but within it the population are classified into a range of groups including 'freemen', 'villeins' and 'slaves'. We can work out that around 10% of the population were 'slaves' who could be bought and sold. We can do this because we know there were 268,984 individuals mentioned in the *Domesday Book* and that 26,898 were 'slaves'. These classifications were based on the individual's labour, assets and relationship to those in power. These classifications are not relevant to a national government today as such categories are now meaningless. Today, the social categories we measure include age, gender and ethnicity. Such measures were not considered a thousand years ago and so were not included in the *Domesday Book*.

More recently, we see that the antecedents of measurement, as we would begin to understand it today, started in the late 18th and early 19th centuries with the emergence of the Industrial Revolution. The shift into Modernity was characterised by a zeal for measuring and classifying everything, including the social (Wright, 1997). In part, the reason for this was down to the economic and sociocultural transformations that were taking place, whereby people left the countryside and headed for the cities in search of work and potentially better opportunities. The social impact of industrialisation also resulted in massive population growth, squalid living conditions, extreme poverty and the rampant spread of disease. To help curtail the many ills of such rapid social change, the early social scientists embarked on a journey of exploration in an attempt to try to explain and thus prevent such things from occurring. In order to explore and explain the social change happening around them, a crucial starting point was measurement; after all, how can you identify and explain a social phenomenon unless you first define what exactly it is that you are trying to identify and explain, whether it be 'race', 'class' or 'poverty'. In part, this was a movement that tried to replicate the natural sciences and apply their thinking to the social world that included testing and measuring statistically.

However, although the Victorians believed themselves to be applying a scientific approach that was value-free and objective, their approach to measurement was culturally biased and reflected their view of the social world. For example, the 19th century was the era of European empire building; hardly surprising then that it was also marked by a mania for measuring race (Thompson, 2006). There were many different attempts to 'measure' race. One example was offered by Georges Cuvier who proposed a three-tier hierarchical classification that positioned those with white skin ('Caucasians') at the top of the intellectual gene pool, those with yellow skin ('Mongolian') in the middle and those with black skin at the bottom ('Ethiopian') (see Mitchell, 2018, for a more detailed account). Here race and intellect were being 'measured' to suppose superior intellect and justify colonial rule. These Victorian racial schema remained widespread until the mid-20th century and were further 'refined' by numerous national governments to justify genocide, ethnic cleansing, apartheid and eugenics. Therefore, it becomes clear when we start measuring human attributes that it is not a value-free process (how can it be?) and can be responsible for the oppression of human beings.

We now know that such racial distinctions are highly prejudiced and discriminatory and so are no longer used in the social sciences or by most governments across the world. Therefore, social research has found more sophisticated ways of measuring such differences, which are now culturally based rather than based on assumptions of intellect and skin tone. Note here, however, the term *race* has been replaced by the term *ethnicity*. Unlike 'race', 'ethnicity' does not have a hierarchy or make any presumptions about intelligence. This is, however, still not a value-free classification, and there is a good deal of cultural baggage incorporated into such a measurement. If not, why would we even bother trying to measure it? Below is the list taken from the UK's ONS (2017a) standard way of collecting information on how people describe their ethnic identity. Does anything stand out?

What is your ethnic group?

Choose one option that best describes your ethnic group or background:

White
1 English/Welsh/Scottish/Northern Irish/British
2 Irish
3 Gypsy or Irish Traveller
4 Any other White background, *please describe*

Mixed/multiple ethnic groups
5 White and Black Caribbean
6 White and Black African
7 White and Asian
8 Any other mixed / multiple ethnic background, *please describe*

Asian/Asian British
9 Indian
10 Pakistani

11 Bangladeshi
12 Chinese
13 Any other Asian background, *please describe*

Black/African/Caribbean/Black British
14 African
15 Caribbean
16 Any other Black / African / Caribbean background, *please describe*

Other ethnic group
17 Arab
18 Any other ethnic group, *please describe*

Box 3.1

Pause for Thought

Before reading on, explore the list above and see if you can identify any anomalies in how this measurement of ethnic grouping is presented?

In answer to the 'Pause for Thought' above, did you question why the category of White was at the top of the list? Why was the list not started alphabetically? Why is it that differences such as Black and Minority Ethnic groups are considered while the notion of 'white' is presented as more unified (Why are differences such as 'Descendant of Eastern European' or 'French' not included?). As previously stated, there is a good deal of cultural baggage in measuring anything within the social sciences. Just as the Victorian measurement of 'race' is no longer appropriate, it is of note that our attempts at measurement throughout the social sciences is in a constant state of flux and liable to change as society changes. Ethnicity is now found in almost all (if not all) social surveys. It would be naive to think that the list above is the finished product and the ONS will keep adapting the list in their attempts to make it more accurate or as trends change and we as a society want to measure our differences in an alternative way.

There are, however, many things that we are still struggling to measure. Take sexual orientation as an example. The ONS acknowledged that the data would be useful and needed by organisations such as the UK's National Health Service but persuaded the government of the time not to put the question in the 2011 Census. There is currently a consultation about whether it will be included in the 2021 Census. If it is included, it will most likely ask about sexual identity rather than sexual orientation with the acknowledgement that such a label does not reflect sexual attraction or sexual behaviour (ONS, 2015c). What do you think is at the heart of this issue of including such questions? One of the issues raised at the time was that of wording complexity. It might be, however, more related to the sensitive nature of the question as it seems we are fine asking about ethnic identity, which seems equally complex.

Controversial and contested measurements

We've considered both sexuality and ethnicity as controversial and contested concepts to measure. Can you think of other similarly controversial concepts to measure? Share your views with a classmate; do they agree? You could work with a partner and each try and develop a different way to measure the same concept; how do your measures differ?

The point here is, however, that when we are trying to measure anything in social research, there are always 'other' things that need to be taken into account. This is never easy, and we need to be aware of how our own sociocultural biases influence how we measure. However, this is not a reason to stop measuring, counting and defining, it could be seen as a societal and research challenge for us to continually strive to get better at measuring things in our social worlds. History will most likely judge our attempts poor, but we cannot get better at something by not doing it. Hence, measure everything, learn something, answer nothing! Importantly, let's keep asking and trying to answer questions because that's what we do!

Box 3.2

Reflective Exercise

In this section of the book, you have been introduced to the complexity of measurement in the social sciences and beyond. Before moving onto the next section, try to create a measure that explores social media addiction. Clearly, there are no correct answers, but there are issues that 'should' be addressed. What would you include?

Units of measurement: variables

Earlier, we discussed different units of measurement for quantifying the physical world, including days and weeks; metres and miles. In social research, we use something called a **variable**. A variable is something (e.g. a characteristic, quantity, attitude or behaviour) that can be measured, including ethnicity, age, income and fear of crime.

Variables are funny things. They can be quite confusing, but this is not the fault of the variable per se but rather the fact that they are being used to try to explain differences in what we see throughout the social world. Think about it, if everyone in the world were the same, we would have no need for variables and social research (or worse, social scientists). The fact is, however, each and every one of us are made up of a vast array of demographic details combined with an assortment of everyday behaviours and beliefs that go to make up what we call variables. That is not to say we do not share categories or labels, such as being 'male' or 'female', 'old' or 'young', 'fat' or 'thin', but combine this with our experiences and we become a complex mix of similarities and differences. This complexity makes social science and descriptive statistics so exciting. At the heart of these endeavours are variables, as they allow us to explore how much we vary from each other. By collecting and interpreting numerical data in the shape of a variable, we can begin to build up a picture of the complexities we see in our societies and across the world. Before we start to examine the complexities of variables, let us first explore how social research uses variables. It might be helpful to break down the process into four main areas:

1   *Classification* – how we define and classify a variable
2   *Counting* – count the number of times a particular classification occurs
3   *Measuring* – the different experiences between those in the classification
4   *Explaining* – offering a possible explanation for the observed differences within a classification

The above four areas are important because if we do not adequately problematise what is meant by a particular classification, we cannot accurately count, measure or explain it. Let us unpack this more and see how we can use the above four areas to help us work with descriptive statistics using **gender** as our example variable.

Classification – gender: men and women?

The first job we must do is to define what is meant by 'gender' and then classify it accordingly. What we are doing is making a variable we are going to call gender. Sometimes this is called 'sex' relating to the biological differences between men and women, yet increasingly, it is being labelled as 'gender', which according to Oakley (1972, 2016) is made up of our societal norms and practices and so is quite different from 'sex'. Sometimes both 'sex' and 'gender' are used interchangeably to mean the same thing. The simplest understanding of 'gender' or 'sex' is that there are two: male and female (although this is increasingly debatable, for instance, see Fisher et al., 2016). The question asked on surveys is relatively simple *'Are you Male or Female?'* The variable would be called *'Sex'* and the choices provided to answer would be 'male' or 'female'. We could get the same answer by asking if the person had XX or XY chromosomes as it is these that differentiate males and females biologically. Doing this, however, might be too confusing for the average person, although more accurate in terms of definitions.

The survey would most likely be asking if the participant was either 'male' or 'female' with them not being allowed to answer 'both'. Yet the gender/sex binary is increasingly being challenged and if the question is not inclusive, it might start to exclude some people from answering. For instance, Trans people might not feel they can answer the question quite as easily as 'male' or 'female'. We therefore might want to include 'Trans male' and 'Trans female'. That would mean that our variable would now have four classifications. An alternative way of asking this is by keeping the 'male/female' binary and including a supplementary question asking if the respondent is transgender ('Yes/No'). In this example, there would be two variables each having two possible answers. Importantly, this is not about defining what it is to be 'male' or 'female' but rather providing the necessary condition to be able to count the number of men and women as a variable so that we can carry out statistical analysis on it either by itself (**univariate analysis**) or with another (**bivariate analysis**). There is not always an easy solution in some areas of the social sciences as many issues are open to debate. For instance, what if someone did not feel they could answer 'male' or 'female' because they did not experience this binary? So when asking about someone's gender, we might provide the following choices from them to select (ordered alphabetically):

Cis male

Cis female

Non-binary

Trans male

Trans female

This is why it is important to consider the classification that you intend to work with before any data is collected or analysed as there might be the need to offer a more sophisticated question and selection of choices to more adequately reflect the social context in which we measure.

Counting gender: how many men and women?

The purpose of classifying a variable is so that the different categories can be **counted** to discover how many, in this instance, men and women are there. Looking at the recorded numbers of men and women in Great Britain between 1938 and 2015

(Table 3.1) shows near equal ratios of men and women with a consistent 1% or 2% more women than men (with the exception of 1940 through to 1942 where there were are 8, 10 and 12 percentage points more women than men). This is administrative data of the type outlined in Chapter 2. Therefore, in the first instance, we would need to count the whole population of the UK and then count how many males or females made up the population.

Table 3.1 UK population between 1937 and 2014

Year	Persons	Males	Females	Males	Females
1937	46,007,600	22,101,900	23,905,700	48%	52%
1938	46,208,100	22,197,100	24,011,000	48%	52%
1939	46,252,700	22,099,700	24,153,000	48%	52%
1940	44,730,200	20,463,100	24,267,100	46%	54%
1941	43,562,400	19,389,200	24,173,200	45%	55%
1942	42,994,000	18,919,400	24,074,600	44%	56%
1943	46,920,600	22,455,200	24,465,400	48%	52%
1944	47,102,600	22,512,700	24,589,900	48%	52%
1945	47,309,900	22,606,500	24,703,400	48%	52%
1946	47,637,800	22,907,400	24,730,400	48%	52%
1947	48,188,700	23,304,200	24,884,500	48%	52%
1948	48,671,200	23,593,100	25,078,100	48%	52%
1949	48,960,000	23,745,900	25,214,100	49%	51%
1950	49,004,500	23,672,100	25,332,300	48%	52%
1951	48,913,900	23,493,600	25,420,300	48%	52%
1952	49,054,200	23,561,000	25,493,200	48%	52%
1953	49,208,800	23,641,900	25,566,900	48%	52%
1954	49,377,600	23,724,400	25,653,300	48%	52%
1955	49,552,300	23,830,200	25,722,100	48%	52%
1956	49,789,000	23,963,000	25,824,000	48%	52%
1957	50,031,700	24,096,200	25,935,500	48%	52%
1958	50,250,200	24,203,500	26,046,700	48%	52%
1959	50,548,600	24,357,500	26,191,100	48%	52%
1960	50,952,700	24,578,700	26,374,000	48%	52%
1961	51,380,036	24,832,293	26,547,743	48%	52%
1962	51,854,828	25,126,167	26,728,661	48%	52%
1963	52,178,200	25,286,436	26,891,764	48%	52%

(Continued)

Table 3.1   (Continued)

Year	Persons	Males	Females	Males	Females
1964	52,532,800	25,479,600	27,053,200	49%	51%
1965	52,881,300	25,651,897	27,229,403	49%	51%
1966	53,167,100	25,792,165	27,374,935	49%	51%
1967	53,470,200	25,947,500	27,522,700	49%	51%
1968	53,710,900	26,051,291	27,659,609	49%	51%
1969	53,946,500	26,169,175	27,777,325	49%	51%
1970	54,104,800	26,244,775	27,860,025	49%	51%
1971	54,387,600	26,412,706	27,974,894	49%	51%
1972	54,557,700	26,501,903	28,055,797	49%	51%
1973	54,692,900	26,575,912	28,116,988	49%	51%
1974	54,708,700	26,593,753	28,114,947	49%	51%
1975	54,702,200	26,607,661	28,094,539	49%	51%
1976	54,692,600	26,606,365	28,086,235	49%	51%
1977	54,666,600	26,591,047	28,075,553	49%	51%
1978	54,654,800	26,575,980	28,078,820	49%	51%
1979	54,711,800	26,618,113	28,093,687	49%	51%
1980	54,796,900	26,656,363	28,140,537	49%	51%
1981	54,814,500	26,654,960	28,159,540	49%	51%
1982	54,746,166	26,606,834	28,139,332	49%	51%
1983	54,765,117	26,612,416	28,152,701	49%	51%
1984	54,852,010	26,659,317	28,192,693	49%	51%
1985	54,988,603	26,723,860	28,264,743	49%	51%
1986	55,110,334	26,773,539	28,336,795	49%	51%
1987	55,222,004	26,826,010	28,395,994	49%	51%
1988	55,331,008	26,877,844	28,453,164	49%	51%
1989	55,486,016	26,952,897	28,533,119	49%	51%
1990	55,641,898	27,040,655	28,601,243	49%	51%
1991	55,831,363	27,125,852	28,705,511	49%	51%
1992	55,961,267	27,184,733	28,776,534	49%	51%
1993	56,078,337	27,241,061	28,837,276	49%	51%
1994	56,218,438	27,306,385	28,912,053	49%	51%
1995	56,375,668	27,399,687	28,975,981	49%	51%
1996	56,502,623	27,476,797	29,025,826	49%	51%
1997	56,642,988	27,555,516	29,087,472	49%	51%
1998	56,797,174	27,639,691	29,157,483	49%	51%

Year	Persons	Males	Females	Males	Females
1999	57,005,421	27,759,976	29,245,445	49%	51%
2000	57,203,121	27,869,972	29,333,149	49%	51%
2001	57,424,178	28,007,991	29,416,187	49%	51%
2002	57,668,143	28,143,761	29,524,382	49%	51%
2003	57,931,738	28,292,214	29,639,524	49%	51%
2004	58,236,322	28,458,718	29,777,604	49%	51%
2005	58,685,543	28,695,680	29,989,863	49%	51%
2006	59,083,954	28,908,469	30,175,485	49%	51%
2007	59,557,392	29,165,626	30,391,766	49%	51%
2008	60,044,620	29,429,625	30,614,995	49%	51%
2009	60,467,153	29,653,650	30,813,503	49%	51%
2010	60,954,623	29,920,958	31,033,665	49%	51%
2011	61,470,827	30,207,937	31,262,890	49%	51%
2012	61,881,396	30,420,674	31,460,722	49%	51%
2013	62,275,929	30,635,876	31,640,053	49%	51%
2014	62,756,254	30,891,041	31,865,213	49%	51%
2015	63,258,413	31,165,316	32,093,097	49%	51%

Note. Adapted from Population Estimates data from UK's Office for National Statistics (accessed 09/08/2017).

As can be seen, counting is not as simple as it first appears, as it is important we also know the population we intend to count. In most instances, it is unlikely that we would be able to count the whole population of Great Britain, so instead, we would try to get a representative sample, in this instance of men and women. By collecting this data, we can start to see if we have equal or near equal ratios of men and women in our sample as we have in the wider population.

Box 3.3

Reflective Exercise

The difference in ratios of men and women between 1940 and 1943 is an interesting deviation from what seems the normal trend. How might you account for this?

Measurement: different experiences of men and women?

Once we know the percentage or ratios of men and women, we can begin to measure if they have different experiences based on their gender. Table 3.1 merely shows the number of men and women in the UK, which is not that interesting in itself; however, the Figure 3.3 compares men and women (the variable *Sex*) in relation to another variable, *'gross weekly earnings'*.

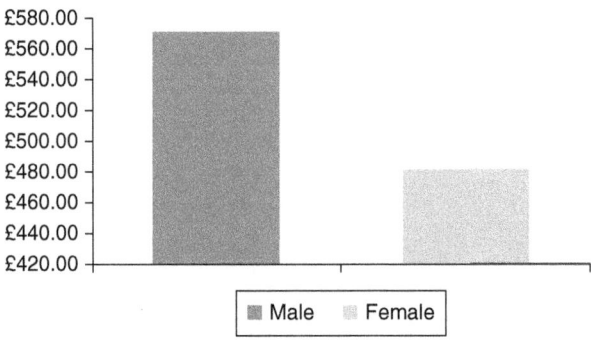

Figure 3.3   Full-time gross weekly earnings by sex

Source. Annual Survey of Hours and Earnings – UK's Office for National Statistics.

Instantly, we can see that men, on average, earn more than women in the UK. Relatedly, we could then compare the UK data with other comparable nations to see if this was a wider phenomenon or a UK-specific problem. Alternatively, we could examine the UK situation in a more sophisticated way by comparing men's and women's pay with their age group included (a new third variable – *'age group'*) to see if the pay gap differs by age group as well as sex (Figure 3.4).

We can see that the pay gap persists at all age groups meaning there is a pattern in pay which we need to explain; we might also note that it appears to widen the older men and women are. We could start by stating how men's pay peaked at the 40 to 49 year group, whereas women's pay peaked at the 30 to 39 year group. We can also state that it is the youngest of both males and females who are paid the least. These important differences would not be known unless we attempted to measure them. As such, we have not offered any possible explanation for such differences, but our identification of these differences could be the starting point of a very interesting data story for us to develop, framed by literature, as we will explore in greater detail in Chapter 7.

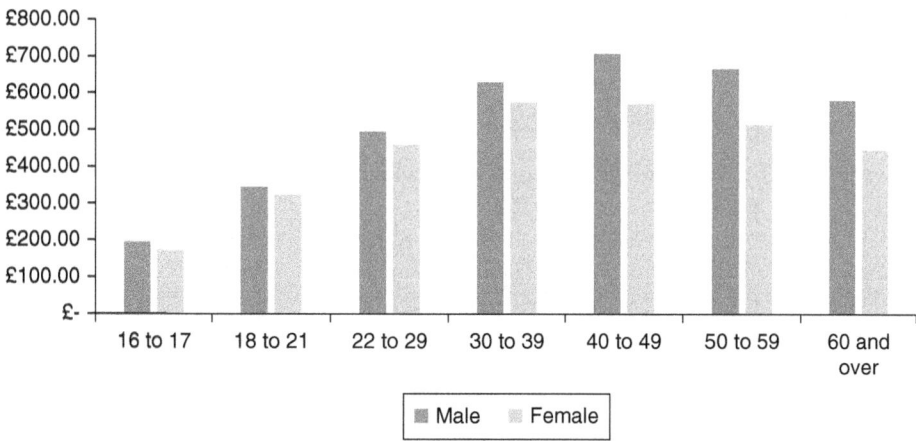

Figure 3.4 Full-time gross weekly earnings by sex and age group

Source. Annual Survey of Hours and Earnings – UK's Office for National Statistics.

Explaining: different experiences of men and women?

The next stage is to explain. As will be discussed in detail in Chapter 7, offering possible explanations as to why social phenomena happen is the key difference between what social researchers do compared to say those who just count up numbers. Remember in Chapter 2 we made the point that most administrative data are collected by people who are not trained researchers and who are not seeking to answer specific research questions; they just want to count up numbers, whether it be patients in hospitals, criminals in prison, or children who passed the national test and so on. The important aspect of explaining is how we provide a narrative of what we think might be occurring. Even our seemingly mundane table (Table 3.1) showing the gender ratio for the UK throws up two interesting patterns. Firstly, the gender ratio remains almost stable (49%:51%) over a century – it would be interesting to compare this with other nations to see whether the gender ratio is naturally occurring and if so, why? Relatedly, we could compare the UK's gender ratio with those nations who have an imbalanced gender ratio (e.g. India and China) to try to explain why. Secondly, in the UK the gender ratio alters between 1940 and 1943, with a much greater number of women to men. This clearly stands out and can be possibly explained by the Second World War. It might be interesting to compare the years 1940–1943 with 1914–1918 to further build an explanation.

Box 3.4

Pause for Thought: Mind the Gap!

Looking at the data above that shows the gender pay gap between men and women of all ages, can you speculate on why this might be. Could it all be down to our biological make up (do men just have more talent, brains and get up and go than women!) or could sociocultural factors (are women just socialised to be homemakers, prioritising family over career, whereas men are expected to earn more as they are the family's breadwinners) also be at play here? You might also look at the variable *'age group'* and ask why this is also important in differences in gross weekly wages. What data-driven stories can you develop to account for these differences?

Levels of measurement

Descriptive statistics focus on identifying and describing variables to provide social insight into different phenomena. Variables, as we have acknowledged, are a means to measure a social attribute and can themselves be placed into different categories depending on how specifically they measure. The Stevens **Levels of Measurement** (1946) is the most widely used approach to categorising variables (see Allen, 2017). Stevens (1946), a psychologist, stated that measurement required two things:

1   That we can categorise every observation.
2   That each observation can be placed in only one category.

If these standards are met, then measurement can be placed at four levels:

1   *Nominal measures.* Any numbers assigned to the categories are arbitrary and not arithmetic; for example, 'gender' may have two categories which we code 1 – male and 2 – female. It would make no difference if we switched the codes around. Similarly, an individual's membership of a category is arbitrary; you are either 'male' or 'female'. Other examples of **nominal variables** would be 'ethnicity', 'nationality' and 'religion'.
2   *Ordinal measures.* Ordinal data can be meaningfully ranked or ordered and can be assigned numbers that reflect this ranking. A classic example of ordinal measurement is a Likert scale, for example, 1 – strongly agree, 2 – agree, 3 – neutral, 4 – disagree, and 5 – strongly disagree. If we switched the codes around, say making 'neutral' 1 and 'agree' 5, then it would affect the analysis. Although ordinal data is ranked, we cannot say that the distance between the ranks is equal or clear.
3   *Interval measures.* Interval data can also be ranked into a meaningful order but in contrast to ordinal data, it is ranked in a clear order, with an equal distance between

categories, for example, temperature. We can do arithmetic on interval data, thus if the temperature is 20 °C and then doubles it would be 40 °C.

4 *Ratio measures.* Ratio measures are the same as interval measures except that they have a meaningful zero point. For example, age, income or population.

However, Stevens's classification has been much amended especially as social science data (which is typically observational) does not conform as readily to it as data from the physical sciences. Therefore, in most textbooks and online, you will usually see a slightly amended version of Stevens's classification, as illustrated in Figure 3.5.

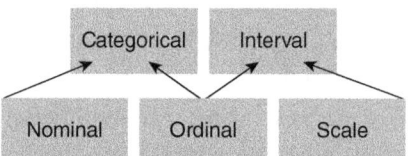

Figure 3.5 Levels of measurement

In this scheme, we can see that Stevens's three original measures (nominal, ordinal and scale) have been kept, but that there is a further level of classification (categorical and interval) at play, which allows us to deal with the ordinal category more effectively. Let's illustrate. Imagine we wanted to measure the variable '*Educational Qualification*'; we could pose the question 'What is your highest educational qualification?' that could have a list of possible categories from 'None' to all the way up to 'PhD'. This list of categories is in a form of ranked order from 'none' to increasingly higher qualifications (you cannot go higher than a PhD). However, there is no equal distance between the categories and an individual's membership of a category is arbitrary. Thus, this form of ordinal measure is much closer to a nominal measure than an interval measure; hence, this scheme would assign it thus, categorical-ordinal. Compare this with a variable seeking to measure life satisfaction; we may pose the question '*How satisfied are you with your life?*' and offer a Likert-type scale ranging from 'Highly satisfied' to 'Highly dissatisfied'. There is a much clearer distance between the categories, albeit membership of each one is still arbitrary. However, because the categories can be ordered, the data is much closer to an interval measure than a nominal measure; hence, this scheme would assign it thus, interval-ordinal. There is a new category, 'scale' (in some text books, this is also called continuous) that is a measure that is identical to Stevens's original interval measure. This is confusing and not aided by different textbooks and software using different labels. For the purpose of this book, we intend to use the names used in IBM SPSS (nominal, ordinal and scale) with the additional distinction of categorical and interval). Let's drill down some more and explore how each of these work in practice.

Categorical variables

Nominal

Nominal variables are useful to distinguish one category from another ('male'/'female') but cannot be ranked or made to represent an exact number (unlike scale that will represent an exact number). A male coded as 1 is not worth half of the female coded at 2! The distinction is only there so that software is able to carry out analysis. Likewise, if we consider hair colour, someone who is blonde (a nominal and categorical variable) is not worth more or less than someone with brown hair. There is no order to the category, meaning that categorical/nominal variables should not and cannot be ordered from high to low. Nominal variables are widely used in descriptive statistics, as they are important for highlighting demographic patterns and key differences. Last, we should recall our earlier discussion regarding how we classify: Nominal variables in particular are subject to sociocultural biases over time, not to mention the subjective preferences and biases of the researcher themselves.

Ordinal

Ordinal variables within the categorical class share many characteristics with nominal variables, including their utility in distinguishing between categories, such as 'highest educational qualification' or 'socio-economic class'. Although they can be placed in a meaningful order (e.g. lowest to highest or highest to lowest), they do not represent an exact number or an equal distance in the ranking. Someone with a PhD coded at 8 is not worth more than someone with no qualifications coded at 1. Such variables are widely used in descriptive statistics.

Interval variables

Ordinal

Ordinal variables are challenging because we can struggle to determine their best 'fit' within this scheme, as they can be either categorical or interval. **Ordinal variables** have a ranked order where the numbers assigned to the categories reflect the ranking. To illustrate, we might use a Likert scale to measure 'attitudes to gender and work', asking the following question: *'Men and women doing the same or similar job should be paid the same?'*

1   Strongly agree
2   Agree

3   Neither agree nor disagree
4   Disagree
5   Strongly disagree

In contrast to a categorical-ordinal variable, the scale could be reversed and meaning would not be lost:

1   Strongly disagree
2   Disagree
3   Neither agree nor disagree
4   Agree
5   Strongly agree

The purpose of the ranking is not intended to demonstrate any particular amount or quantity but is there to differentiate between the different degrees of agreement or disagreement. What is important, however, is that they have similar characteristics to a scale variable (see below) and can often be treated as a scale variable in statistical analysis.

Scale

Scale variables measure quantifiable distances between categories that are on a scale where each category is an equal distance from the last; the distance between 11 years and 12 years is broadly the same as that between 22 years and 23 years. We know that someone who is 2 years older than someone aged 30 will be 32 years. However, if we wanted to explore how long someone spent using the internet, or watching television, it would be appropriate to use another unit of measurement, mostly likely the hour. That would mean that someone spending 20 hours a week on the internet is spending twice as long as someone who uses it for 10 hours per week. The scale measure used to count them is both quantifiable and comparable. It can still become confusing when we start to count what seems like nominal data, for instance, the number of friends someone has on Instagram or Facebook. However, this would still be classed as scale data as you are counting the number of friends on your social network site. Someone who has 100 friends has half the number of friends than someone who has 200. It is not possible to distinguish someone who is your best friend on Facebook from someone who is more of an acquaintance (they can be a bad friend, but they cannot be half a friend!). Here it would be necessary to use a nominal classification of 'friends' and distinguish them from 'acquaintance'. The scale data would be the number of friends or acquaintances. It is important to note that not all numbers are scale variables – student ID or National Insurance numbers are nominal variables as they cannot be added, subtracted, and so on. To think of it in another way, a soccer player wearing a Number 10 shirt is not worth 10 times that of a goal-keeper wearing a Number 1 shirt or half the amount of someone with the Number 20 shirt. Scale variables are widely used in inferential analysis as they are numeric and therefore more complex calculations can be done.

Box 3.5

Time to Get Your Hands Dirty! Identifying Variable Types

The final part of this chapter is an activity to help develop your understanding of how we measure things in the social sciences. We have put the levels of measure diagram here to assist you.

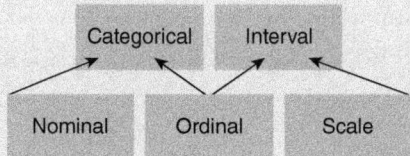

Figure 3.5 Levels of measurement

Below is a short vignette that tells the story of James. Can you identify all the different variables and then the variable types? We estimate there are 20 but you might find more.

James's Story

James is 24 and lives in the North of England. He left school at 16 and trained to be a mechanic. He is engaged to be married to Michelle. He still lives with his parents and is saving up to buy a house with Michelle. He doesn't smoke but likes the occasional drink at the weekend. His favourite soccer team is Tranmere Rovers, and he is a season ticket holder. He also plays football for the local five-a-side team and volunteers as a football coach for the local youth club. James is a committed vegetarian, so has a plant-based diet. He doesn't follow any recognised religion but does consider himself to be spiritual. He has never voted previously and doesn't consider himself political, but in the last election, he voted for Jeremy Corbyn.

Task 1

– Identify all the variables in the above case study by underlining them.

Task 2

– How would you label the variables? (nominal, ordinal, scale)

The answers to this can be found in Table 3.2 if you want to check.

Table 3.2 Variables identified from James's story

Variable	In Vignette	Variable Type
Age	24 years old	Scale
Location where lives	North of England	Nominal
Age left school	16 years old	Scale
Occupation	Trained mechanic	Nominal
Relationship status	Engaged	Nominal
Sexual orientation	Heterosexual	Nominal
Homeowner status	With parents	Nominal
Financial status	Saver	Nominal
Health habits	Non-smoker	Ordinal or nominal
Drinking alcohol	Light drinker	Ordinal or nominal
Football club supported	Tranmere Rovers	Nominal
Football Club supporter	Season ticket holder	Ordinal or nominal
Football player	5-a-side	Ordinal or nominal
Football coaching	Local youth club	Nominal
Vegetarian/carnivore	Vegetarian	Nominal
Religiosity	Does not follow any religion	Nominal
Religiosity	Spiritual	Nominal
Voting practices	First time voter	Ordinal or nominal
Political affiliation	Non-political	Nominal
Party politics	Voted Jeremy Corbyn	Nominal

Looking ahead

This chapter has dealt with the complexity of measurement. In the physical world, we might think that measurements such as height, weight and time are easy. The reality, however, is very different as all measurements have a cultural element embedded within. When we start to measure things in our social world, they are no more or less easy or hard. We just have to think about them more, not least to ensure that if we are measuring things attributed to people, we do not discriminate against them or use crass stereotypes to define them. In Chapter 4, we will develop the different ways that categorical variables (both nominal and ordinal) can be used in descriptive statistical analysis. Here, we introduce how the social sciences can work with these types of variables using something you will likely be very familiar with: percentages. However, don't be fooled that you already know about percentages because you did them at school. What we will focus on is how they can be used when working with large amounts of data to help explain the trends and patterns when using just one variable (univariate analysis) or two variables (bivariate analysis).

Chapter Summary

- The point of measuring anything is so that we have information that is useful to us.
- Validity is an important concept in social research and is linked to the concept of reliability, which relates to the effectiveness of a tool of measurement.
- With social research, it is more difficult to achieve precise measurements due to the complexity of what is being measured.
- Measuring human attributes is not a value-free process and can be responsible for the oppression of human beings. There is a good deal of cultural baggage in measuring anything within the social sciences. Our attempts at measurement is in a constant state of flux and liable to change as society changes.
- Each and every one of us are made up of a vast array of demographic details combined with an assortment of everyday behaviours and beliefs that go to make up what we call variables.
- Social research uses variables in four main ways: *Classification, Counting, Measuring* and *Explaining*.
- The Stevens Classification (1946) is the most widely used approach to categorising variables and has been adapted to be classified as categorical and interval measurements, which encompass nominal, ordinal and scale measurements.

Further Reading

Chapter 3 focused on measurement, if you would like to delve deeper into issues of measuring social concepts, then try some of the following readings.

Mosher, C. J., Miethe, T. D., & Hart, T. C. (2011). *The mismeasure of crime*. Sage.

This book is really helpful in showing the complexities involved with measuring anything in the social world, but specifically, the measurement or mismeasurement of crime. The media often sensationalise crime, but in this book the authors cut through all the lurid headlines and misuse of statistics to reveal how data can be presented in a way that does not enlighten but rather misleads the reader.

Connelly, R., Gayle, V., & Lambert, P. S. (2016). Ethnicity and ethnic group measures in social survey research. *Methodological Innovations, 9*, 1–10 https://doi.org/10.1177/2059799116642885

Describing ethnicity as a fuzzy concept, Connelly et al. (2016) captures the complexity and possible acrimony of such measurements. What is useful about this article is how it is aimed at the non-expert researcher wanting to work with ethnicity when using secondary data. In the work, these authors examine two standardised approaches often used in survey work termed as the 'mutually exclusive category approach', which focuses primarily on skin tone and national origins, and the 'multiple characteristics approach', which offers a more detailed way of examining the different aspects of ethnicity. In doing this, the authors reveal that in spite of our best efforts, the measurement of ethnicity is far from complete.

Forbes, A. (2017). Measuring gender identity. In K. Nadal (Ed.), *The SAGE encyclopedia of psychology and gender* (pp. 1133–1136). Sage. https://doi.org/10.4135/9781483384269.n379

Forbes (2017), in *The SAGE Encyclopedia of Psychology and Gender*, offers alternative approaches to the measurement of gender (specifically trait measures and ideology measures) that she contends can better capture the experience of gender. This is useful in that she introduces the complexities that arise when moving away from binary measurements of cis/trans male/female.

Westbrook, L., & Saperstein, A. (2015). New categories are not enough: Rethinking the measurement of sex and gender in social surveys. *Gender & Society, 29*(4), 534–560. https://doi.org/10.1177/0891243215584758

Westbrook and Saperstein's (2015) article is an interesting exploration of the social constructions of gender and the difficulties that have arisen when measuring this construct. By taking a systematic approach to how sex/gender has been measured in social science surveys, they are able to show how the categories of 'male' and 'female' cannot cover all possible ways of being and argue for a move away from the cisnormativity that permeates many survey gender/sex questions.

4

I AM NOT A NUMBER, I AM A CATEGORICAL VARIABLE

Chapter Overview

Introduction 82

Percentages: a story of parts and wholes 85

Categorical data and percentages 87

Working with valid percent and percent: working with missing data 95

Writing up results: producing descriptive summaries 99

What to report with missing data? 101

Strengths of using the percent 103

Working with contingency tables 105

How to guides for IBM SPSS and MS Excel? 107

How to guide for IBM SPSS? 107

How to guide for MS Excel 115

Looking ahead 127

Further Reading 128

Introduction

In Chapter 3 we explored measurement, so hopefully you have some knowledge and understanding of the purpose and challenges of measurement particularly in social science research. You also learnt that there are different types of variables that allow us to measure different types of social phenomena. In this chapter, we build on this learning to explore a particular type of variable and therefore data: categorical. In Chapter 1, we discussed briefly the most common ways that categorical data is analysed using descriptive statistics. This chapter develops this further by looking in more detail at univariate (using just one variable) and bivariate (using two variables) analysis to show the strengths (and some limitations) of categorical data at the descriptive level of statistical analysis.

Throughout the chapter, we will be introducing IBM SPSS and MS Excel 'how to' screenshots that you can use to help you learn how to carry out both univariate and bivariate analysis of categorical data. One thing to note is how IBM SPSS and MS Excel work very differently with the data. This will be explained as you work through the chapter. Obviously, there are many other statistical software available, including free open-source software like R; you can easily find online guides to using these software or you could use a specialist textbook. You can, of course, analyse categorical data manually especially if you have small samples and you can find easily online the key equations and calculations needed to do this. We would advise using a statistical software to analyse categorical data from large samples.

As highlighted in Chapter 1, the purpose of this volume of the series is to focus on descriptive and exploratory statistics. To this end, we will separate categorical measurement/data from interval measurement/data as they are quite different in what we can do with them statistically. Figure 4.1 shows the types of variables that are being introduced here.

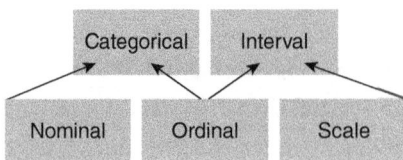

Figure 4.1 Types of variables

This chapter will focus on categorical variables that includes both nominal and ordinal data. It is OK if you have jumped straight to this chapter as long as you have a basic understanding of the differences between categorical and interval data. If you're not sure, why not read Chapter 3 pages 72–76 that will provide you with the basics needed to continue with this chapter.

A brief recap from Chapter 3 shows that categorical data can come in two forms:

- Categorical (nominal)
- Categorical (ordinal)

While these can both be categorical variables, they are in fact very different. Nominal variables are things like 'gender', 'ethnicity', 'place of birth', 'favourite ice-cream flavour' and so on. As such, a key feature of categorical (nominal) data is that it has two or more categories but that they have *no* order or structure to them. To illustrate using the 'favourite ice-cream flavour' example; you are either 'vanilla', 'strawberry' or 'chocolate', it is an arbitrary categorisation based on your taste preference, with no order to the flavours or hierarchy of taste. The categories in nominal data are often referred to as 'groups' or 'levels'; you can use any of these three terms and it would be correct, but just remember to consistently use the same one throughout your work. This type of data would often be known as nominal data and is usually the type of data that most students find easy to recognise and work with. However, nominal data has a difficult job to do, as it involves the not so simple act of giving a name to a category or group of people. This could be the category of 'male' or 'female', if we were being binary about categorising people by their 'gender'. It could be more complicated (and inclusive?) and include 'Cis-Male', 'Cis-Female', 'Trans-Male', 'Trans-Female' and 'Non-Binary'. These are just categories that you could be fitted into. In the example above of gender, it would only be possible to fit in one (or no) category. As such, this is very much the counting stage of working with quantitative data in that once the measurement has been agreed upon (in this case the respondent's 'gender') it would be a matter of then counting the numbers of people who fall into each of the variable's categories, whether it be two or five categories. That is not to say the categories cannot or will not change. They represent the current zeitgeist of what is considered to be our 'gender'. Perhaps the focus also needs to be on the time when the question is being answered in that if at that time you identify as a 'man' or 'woman', 'trans or cis', then this is the answer you can give. The fact that you might identify differently the next day does not matter as it is on the day of the data collection that it matters, and on that day, you belonged to just one category.

The important point is how it can be possible to belong to only one category. For instance, if you were born in Lancashire, UK, it is not also possible to be born in Yorkshire too (unless you were born while travelling across the Lancashire/Yorkshire border, something that does not seem that likely!). Of course, we live in ever-changing times, and as indicated in Chapter 3, such things as 'gender' and 'ethnicity' are much more fluid today and therefore under constant review.

Ordinal variables are similar to nominal with a key difference being that the categories can be ordered in some way. The keen eyed amongst you will have noticed that ordinal data is in fact included in both the categorical and interval variable types. We will explore the interval-ordinal variables in Chapter 5 as it is better to get to grips with categorical-ordinal variables before adding another layer of complexity. The key to ordinal data is that the classification can be ordered and ranked.

Let's take the example of qualifications. This data could be ordered from 'lowest qualification' or 'no qualifications', to the highest which would be a 'PhD'. We know, for instance, that in the UK context A-levels are a higher level qualification than GCSEs. Similarly, we know that an MSc degree is a higher level than a BSc degree. The subject does not matter with this hierarchy, it is the level at which the learning has taken place (BSc is studied before an MSc). It is in this way that they are ordered. Yet in the same way as income levels can categorise people into different groups, so can qualifications. For the UK context, our groups or categories within our variable could be as follows:

1   No formal qualifications
2   GCSE
3   A-level
4   Bachelor's degree
5   Master's degree
6   PhD

Can you see how they are both categories of people (those with no qualifications, those with GCSEs, those with MSc degrees etc.) while also having some order to the categories. An A-level qualification is higher than a GCSE, but lower than a bachelor's degree. There is no precise distance between each qualification level; they are simply in an ascending order. Ordinal data can also be attitudinal such as a Likert Scale. For example, a survey question could ask how much you agree or disagree with the following statement, '*Working with quantitative evidence is difficult*':

5   Strongly agree
4   Agree
3   Neither agree nor disagree
2   Disagree
1   Strongly disagree

Again, it is easy to see that there is an order to the categories, but as highlighted in Chapter 3, this is not a precise measurement, such as length, time or weight. Someone who 'agrees' with the statement might have similar sentiments to someone who 'strongly agrees'; we have no way of knowing as we are asking a subjective question. A similar type of question could be asked about how anxious someone feels. The ONS, for example, have a question asking,

On a scale where 0 is 'not at all anxious' and 10 is 'completely anxious', overall, how anxious did you feel yesterday?

It might be that you had an exam yesterday so felt somewhat anxious and gave an answer of 8. Another person might be attending a new class and be quite a shy person. They too could give an answer of 8. The issue here is that their 8 and your 8, while the same score on an anxiety scale, might experientially, be very different. As anxiety has no precise measure in the same way as hours or weight do, the point is to measure an overall feeling for anxiety. Interval-ordinal variables will be explored in much more detail in Chapter 5.

Percentages: a story of parts and wholes

When we analyse categorical data, we are interested literally in the number (called frequency) of respondents in each category of a specific variable, whether it be 'sex', 'ethnicity', or 'social class'. This is a simple process of counting the numbers of each category to give us the 'count' or 'frequency'. With small samples, you can do this manually, but with large samples, you will want to use statistical software; later in this chapter, we will show you how to do this with IBM SPSS and MS Excel. However, a simple count per category is not that meaningful; we want to get a sense of how it relates to the other categories within the variable. To do this, we need to use percentages, yes, those things you learnt in primary school; we did say that descriptive statistics had a joyful simplicity to it. Percentages are really useful as they offer a standardised way to examine, interpret and report data.

Percent literally means 'out of 100'. With 15%, we know that this is 15 out of 100 or 15 parts out of 100. There is a common currency with percentages and they are something that are used on a daily basis. For example, if you are in a class of 100 and there are 35 males and 65 females, that means there are 35% males and 65% females. It is relatively simple when we are dealing with groups of 100, although it can become more complicated with different group sizes. Yet with a very simple calculation, it is possible to work out the percentage of any group size. All that is needed is to:

1 divide the part (the part of the group you want to find)
2 by the whole group number
3 then multiply it by 100.

To illustrate, let's say we asked 50 students to identify their main dietary preference and 15 students identified as 'vegan'. If we wanted to work out what percentage of our students is 'vegan' we would do the following calculation:

$$\left(\frac{15}{50}\right) \times 100 = 30\%$$

This is why using percentages is a story of parts and wholes. Both are needed to be able to calculate the percentage. You can calculate percentages manually, with a calculator or let your statistical software do the work.

Box 4.1

2-Minute Recap!

Set a timer on your phone for 2 minutes and answer the following:

In a recent social science lecture of 265 students, there are 97 males and 168 females. What percent of the class are female and what percent of the class are male? Before finding the answer, just remember the following:

- The whole is the number of students in a social science lecture = 265.

- There are two parts (males = 97 and females = 168). We want to find out the percent of females and then percent of males attending the lecture.

The very idea of percent has quite a mixed heritage stemming back to the Latin name ('per' meaning 'out of' and 'cent' meaning '100'). But it was the Greeks who originally saw the value of considering parts of 100. It was not until the 15th century that the percent symbol that we know today appeared. You might be asking why percentages and their history is important. Well, their importance most likely results from how they have endured for such a long time, and this is linked to how easily they help us explain and understand numbers.

Box 4.2

2-Minute Recap!

Set the timer on your phone for 2 minutes and then calculate the percentage of the following:

Four hundred and sixty-eight 21-year-olds were asked if they had passed their driving test: 56 females and 52 males had. What percent of females and males had passed their driving test?

Males =

Females =

Make it competitive by challenging a classmate and comparing times!

Categorical data and percentages

As described earlier, the type of variable, be it nominal, ordinal or scale, will determine what statistical technique is used. For percentages, the type of variable used would be categorical, either nominal or ordinal. In Chapter 1, we highlighted why descriptive statistics are so useful due to them being able to summarise vast amounts of data on what are called frequency tables. A **frequency table** is quite simply a table that shows the frequency that something occurs in a data set. It might be the number of males or females included in a data set or the different ethnic groupings that are being measured. By producing frequency tables for categorical data, we can quickly interpret the data and 'raw' conclusions in a way that would not be possible by simply looking at the 'raw' data. To illustrate this, we have created some fictional data (*Ice cream dataset 2019*). Figure 4.2 shows the top of the data set in 'Data View' on IBM SPSS. As can be seen in Figure 4.3, which is taken from the bottom of the data set, there are 1251 respondents who took part in this particular survey. By just looking at the top and bottom of the IBM SPSS Data view, think for a moment, how you could summarise the vast amount of data.

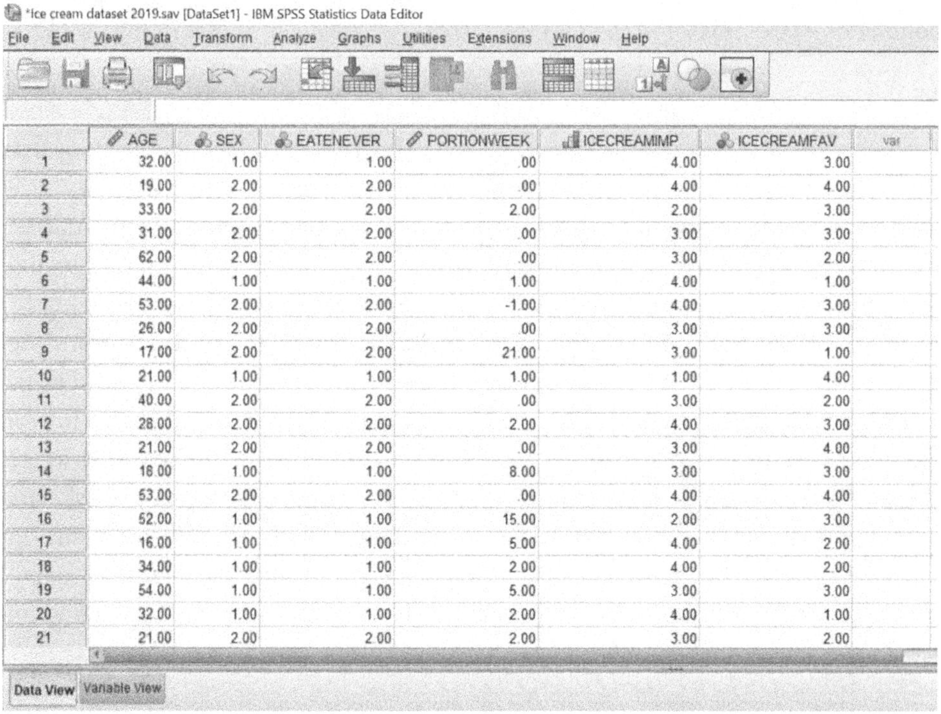

	AGE	SEX	EATENEVER	PORTIONWEEK	ICECREAMIMP	ICECREAMFAV	var
1	32.00	1.00	1.00	.00	4.00	3.00	
2	19.00	2.00	2.00	.00	4.00	4.00	
3	33.00	2.00	2.00	2.00	2.00	3.00	
4	31.00	2.00	2.00	.00	3.00	3.00	
5	62.00	2.00	2.00	.00	3.00	2.00	
6	44.00	1.00	1.00	1.00	4.00	1.00	
7	53.00	2.00	2.00	-1.00	4.00	3.00	
8	26.00	2.00	2.00	.00	3.00	3.00	
9	17.00	2.00	2.00	21.00	3.00	1.00	
10	21.00	1.00	1.00	1.00	1.00	4.00	
11	40.00	2.00	2.00	.00	3.00	2.00	
12	28.00	2.00	2.00	2.00	4.00	3.00	
13	21.00	2.00	2.00	.00	3.00	4.00	
14	18.00	1.00	1.00	8.00	3.00	3.00	
15	53.00	2.00	2.00	.00	4.00	4.00	
16	52.00	1.00	1.00	15.00	2.00	3.00	
17	16.00	1.00	1.00	5.00	4.00	2.00	
18	34.00	1.00	1.00	2.00	4.00	2.00	
19	54.00	1.00	1.00	5.00	3.00	3.00	
20	32.00	1.00	1.00	2.00	4.00	1.00	
21	21.00	2.00	2.00	2.00	3.00	2.00	

Figure 4.2 'Raw' data taken from the Ice cream data set 2019 (IBM SPSS Data View, top)

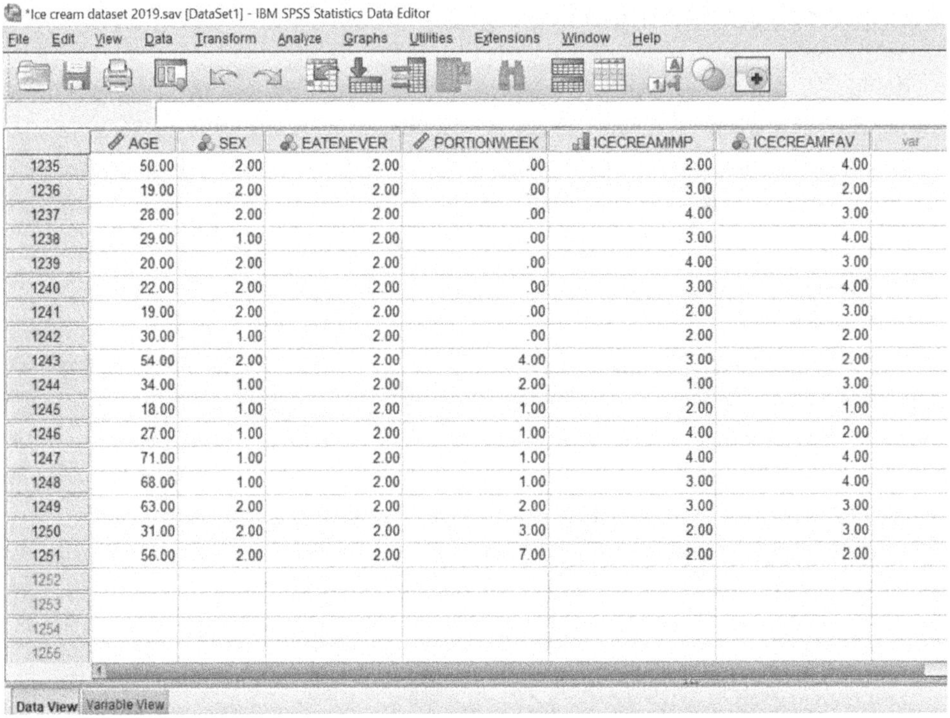

	AGE	SEX	EATENEVER	PORTIONWEEK	ICECREAMIMP	ICECREAMFAV	var
1235	50.00	2.00	2.00	.00	2.00	4.00	
1236	19.00	2.00	2.00	.00	3.00	2.00	
1237	28.00	2.00	2.00	.00	4.00	3.00	
1238	29.00	1.00	2.00	.00	3.00	4.00	
1239	20.00	2.00	2.00	.00	4.00	3.00	
1240	22.00	2.00	2.00	.00	3.00	4.00	
1241	19.00	2.00	2.00	.00	2.00	3.00	
1242	30.00	1.00	2.00	.00	2.00	2.00	
1243	54.00	2.00	2.00	4.00	3.00	2.00	
1244	34.00	1.00	2.00	2.00	1.00	3.00	
1245	18.00	1.00	2.00	1.00	2.00	1.00	
1246	27.00	1.00	2.00	1.00	4.00	2.00	
1247	71.00	1.00	2.00	1.00	4.00	4.00	
1248	68.00	1.00	2.00	1.00	3.00	4.00	
1249	63.00	2.00	2.00	2.00	3.00	3.00	
1250	31.00	2.00	2.00	3.00	2.00	3.00	
1251	56.00	2.00	2.00	7.00	2.00	2.00	
1252							
1253							
1254							
1255							

Figure 4.3 'Raw' data taken from the Ice cream data set 2019 (IBM SPSS Data View, bottom)

The six variables shown in Figure 4.3 data set have such vast amounts of data, it would be impossible to provide any useful information about the data other than the variable name (and this is just for six variables when an actual data set has many, many more). The above data is shown in IBM SPSS. It should be noted that this software package only works with numerical data such as presented in Figure 4.3. You might be wondering what all the different numbers mean. Well for age, this is easy as this is a scale variable (see Chapter 5). The others are in fact categorical variables. By just looking at the numbers, it would be difficult to work out the different possible answers available for respondents to choose. However, these different choices are called 'Values' in IBM SPSS, which can be associated with the text values of the different answers. This will become clearer at the end of the chapter in the 'How to' section. Before we get to that, see Figure 4.4, which provides the information for each variable starting with the variable name and label followed by the values.

Adult number 1 (respondent): *SEX* (this is the variable name and label)

1. = Male

2. = Female

Variable Values and associated names

Adult number 1 (respondent): *AGE* – (no predetermined value provided for Scale variables as actual number used). ?

Respondent ever eaten ice cream, *EATENEVER*:

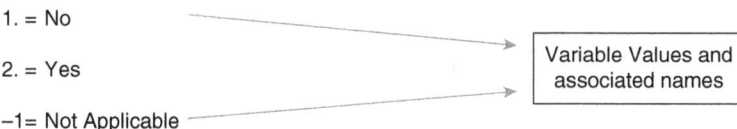

1. = No

2. = Yes

–1= Not Applicable

Variable Values and associated names

Number of portions of ice cream consumed per week, *PORTIONWEEK*: – (no predetermined value provided for Scale variables as actual number used).

Ice cream is an important part of my diet, *ICECREAMIMP*:

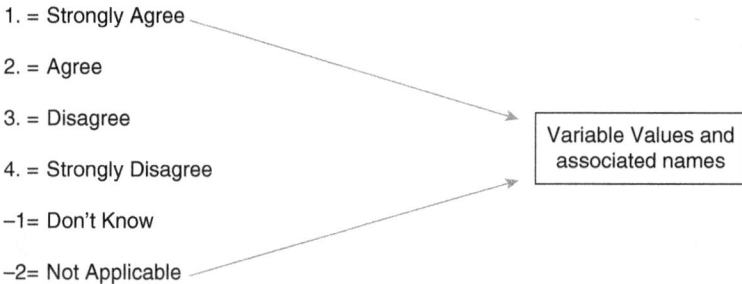

1. = Strongly Agree

2. = Agree

3. = Disagree

4. = Strongly Disagree

–1= Don't Know

–2= Not Applicable

Variable Values and associated names

What is your favourite flavour of ice cream, *ICECREAMFAV*:

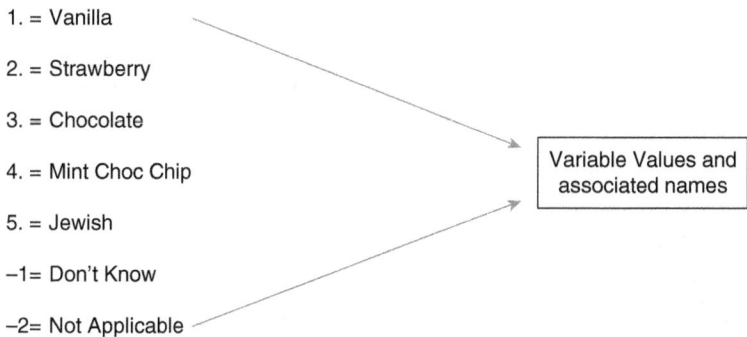

1. = Vanilla

2. = Strawberry

3. = Chocolate

4. = Mint Choc Chip

5. = Jewish

–1= Don't Know

–2= Not Applicable

Variable Values and associated names

Figure 4.4   Information for each variable

With this additional information, it is possible to begin to combine the numerical data with the text answer. Figure 4.5 is the same table that shows the different values that respondents can choose in IBM SPSS.

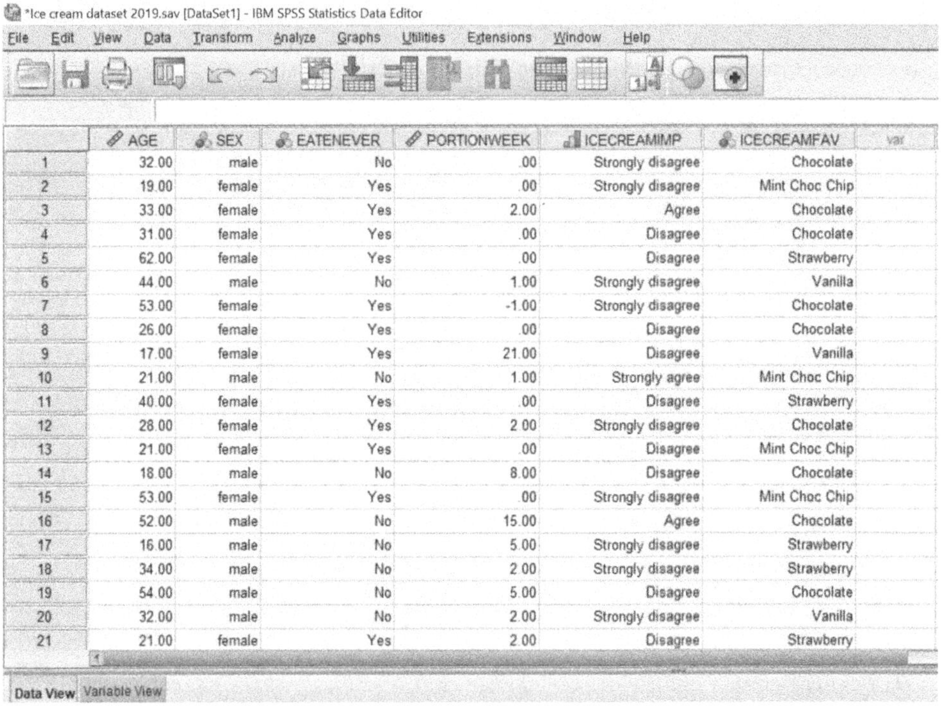

*Ice cream dataset 2019.sav [DataSet1] - IBM SPSS Statistics Data Editor

File   Edit   View   Data   Transform   Analyze   Graphs   Utilities   Extensions   Window   Help

	AGE	SEX	EATENEVER	PORTIONWEEK	ICECREAMIMP	ICECREAMFAV	var
1	32.00	male	No	.00	Strongly disagree	Chocolate	
2	19.00	female	Yes	.00	Strongly disagree	Mint Choc Chip	
3	33.00	female	Yes	2.00	Agree	Chocolate	
4	31.00	female	Yes	.00	Disagree	Chocolate	
5	62.00	female	Yes	.00	Disagree	Strawberry	
6	44.00	male	No	1.00	Strongly disagree	Vanilla	
7	53.00	female	Yes	-1.00	Strongly disagree	Chocolate	
8	26.00	female	Yes	.00	Disagree	Chocolate	
9	17.00	female	Yes	21.00	Disagree	Vanilla	
10	21.00	male	No	1.00	Strongly agree	Mint Choc Chip	
11	40.00	female	Yes	.00	Disagree	Strawberry	
12	28.00	female	Yes	2.00	Strongly disagree	Chocolate	
13	21.00	female	Yes	.00	Disagree	Mint Choc Chip	
14	18.00	male	No	8.00	Disagree	Chocolate	
15	53.00	female	Yes	.00	Strongly disagree	Mint Choc Chip	
16	52.00	male	No	15.00	Agree	Chocolate	
17	16.00	male	No	5.00	Strongly disagree	Strawberry	
18	34.00	male	No	2.00	Strongly disagree	Strawberry	
19	54.00	male	No	5.00	Disagree	Chocolate	
20	32.00	male	No	2.00	Strongly disagree	Vanilla	
21	21.00	female	Yes	2.00	Disagree	Strawberry	

Data View   Variable View

Figure 4.5   'Raw' data taken from the Ice cream data set 2019 (IBM SPSS Data View, text)

As can be seen, there is still no actual analysis of the data itself. IBM SPSS has a toggle button (see Figure 4.6) that allows the user to switch between the numeric and text value. Note that this can only be accessed in Data View.

To make any sense of this data, we need to generate some frequency tables that will summarise all this data, showing us how often something occurs. This time we will use an actual data set, the CSEW 2016–2017. Let's start by looking at a frequency table (using IBM SPSS) for the categorical variable *sex*. Table 4.1 shows tables from IBM SPSS called an 'Output'. It is the result from whatever statistical analysis has been performed in IBM SPSS. In this instance, they are frequency tables but they could be graphs or charts depending on which best summarises the data.

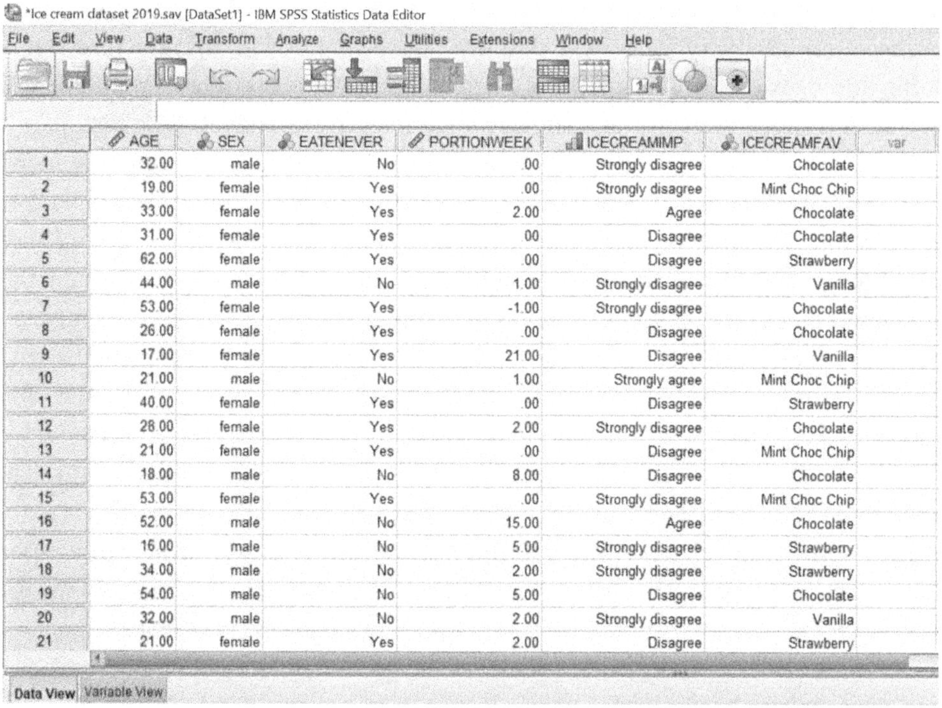

Figure 4.6 'raw' data taken from the Ice cream data set 2019 (IBM SPSS Data View, showing text – toggle button)

Table 4.1 IBM SPSS frequency table for sex taken from the CSEW 2016–2017

Frequencies

Statistics

Adult number 1 (respondent): Sex

N	Valid	35420
	Missing	0

Adult number 1 (respondent): Sex					
		Frequency	Percent	Valid Percent	Cumulative Percent
Valid	Male	16437	46.4	46.4	46.4
	Female	18983	53.6	53.6	100.0
	Total	35420	100.0	100.0	

Note. CSEW = Crime Survey of England and Wales.

If Table 4.1 is compared with Figures 4.4, 4.5 and 4.6 shown previously, it becomes immediately clear why we need to produce frequency tables so that the data can be summarised, examined and analysed. The first part of the output table that IBM SPSS produces is called 'Statistics' and shows the number of respondents who took part in research (who are known as the respondents). In this instance, the sex of respondents is shown followed by the number of people who did not answer the question (known as 'Missing'). The first thing of note is how there are 35,420 respondents (a large sample of respondents) who answered this question leaving no one who did not answer.

The second output table that is produced is the actual frequency table. As there are no missing cases, the total, found at the bottom left-hand side, should have the same number of respondents answering the question (this is the case as again the table shows 35,420). Don't take our word for it, check to see if the figures are the same.

What is also shown are the counts of males and females along with the data expressed as a percentage (we will deal with valid and cumulative percent later in the chapter). Looking at the data alone would not allow us to fully scrutinise and report on the data to the same degree as we can when we use a frequency table. By using a frequency table, we can immediately see that there are 46.4% ($n = 16,437$) men and 53.6% ($n = 18,983$) women in the sample. The percentage should add up to 100%, and the number of males and females – if there is no missing data – should add up to that reported in the first table.

Box 4.3

2-Minute Recap!

Set a timer on your phone for two minutes and do the following:

Check the frequencies of males and females correspond to the total.

Check the percent of males and females add up to 100%.

When faced with a frequency table, it is easy to quickly glance at it and hope to see all the necessary aspects. However, the above table only has a few details. Sometimes, the frequency table can have many different parts of the findings that need to be explored. As such, it is important that you begin to learn how to 'read' the table. In this instance, we take 'read' to mean 'comprehend, decipher or to simply make sense of' the data. This is called **data familiarisation** and is an essential part of EDA. Remember, EDA in part is about summarising the main characteristics of frequency tables. It therefore requires you to make informed decisions about not only what can be included but also what can be ignored. This might not be as easy as it sounds at first, as in essence, you are perhaps doing something you have never attempted before – to read a frequency table. The mastery you have with spoken and written language comes from years and years of practice and tuition. So, if the frequency tables do look a little overwhelming, don't be too harsh on yourself as this is to be expected. But don't give up! Keep working through them and keep asking 'what does the data tell us?' Then look at all the different parts of the table and ask if it is important, does it add to what you are exploring. Quite often, it is all important in some way because it all helps us understand the data being examined. However, there might be things in there like little additional pieces of information, such as 'Adult number 1 (Respondent): Sex', don't let this distract you. This is just the name and the label of the variable. In short, it is important not to give up or to rush. Instead, practice, practice and practice some more. They will become easier to read, honest.

Let's look at what the data in the frequency table for *Sex* is telling us. Well, the first thing of note is that there are more women who took part in this survey compared with men. This is important information if the data is to be trusted as it looks like it broadly represents the percentages of males and females who live in England and Wales (the countries where this survey took place). If there were vastly different percentages of men and women, we would need to question as to why this might be. We might even question if we can trust the data set.

Box 4.4

2-Minute Recap!

Set a 2-minute timer on your phone and answer the following questions.

1   What is the percentage of women in the frequency table?
2   What is the percentage of men in the frequency table?
3   How many more women are there to men?
4   How many respondents did not answer the question ('asking them what was their sex')?
5   What is the cumulative percentage for both men and women?

Moving on, let's examine the variable for '*yrsarea*'. You might be wondering why it has such a strange looking name. Well, the reason for this is because with IBM SPSS, the variable NAME cannot have any spaces or special characters which is why they are not as informative as they could be. To help them make sense, the variable LABEL is also provided so that the necessary information is there to help identify what the variable is measuring. In the example below, the name of the variable is *'yrsarea'* and the label is *'How long lived in area'*. Frequency tables for IBM SPSS will display both but you might find it useful to keep to the default and display the variable label. This is shown in Table 4.2 – the output table.

Table 4.2 IBM SPSS output frequency table for *yrsarea* (How long respondents have lived in the area)

⇨ **Frequencies**

Statistics

How long lived in this area

N	Valid	35413
	Missing	7

How long lived in this area

		Frequency	Percent	Valid Percent	Cumulative Percent
Valid	Less than 12 months	2125	6.0	6.0	6.0
	12 months but less than 2 years	1946	5.5	5.5	11.5
	2 years but less than 3 years	1990	5.6	5.6	17.1
	3 years but less than 5 years	2662	7.5	7.5	24.6
	5 years but less than 10 years	4509	12.7	12.7	37.4
	10 years but less than 20 years	6900	19.5	19.5	56.8
	20 years or longer	15281	43.1	43.2	100.0
	Total	35413	100.0	100.0	
Missing	Refused	5	.0		
	Don't know	2	.0		
	Total	7	.0		
Total		35420	100.0		

There is a lot more to examine in this output table displaying *'yrsarea'* because there are more choices (seven in total) for respondents from which to choose. Again, the first table that IBM SPSS produces is a count of those who answered the question along with any who did not (the 'Missing'). Can you see how in this example, there are seven respondents who did not answer the question? Examining the second table, you will see that the seven comprise of five who 'refused' to answer and two who stated they 'don't know'. That leaves a total of 35,413 respondents who answered this particular question. As with the *'Sex'* frequency table output, there is again the count for each choice along with the percentage of respondents in each category. This time however, the frequency table provides two totals: one that includes the missing (*n* = 35,420) and one that excludes those who did not answer the question (*n* = 35,413).

If you are able to see the two totals in their respective locations on the table, then you have begun to learn the important skill of being able to read frequency tables.

The question is however, 'Why are two totals provided?' It can be confusing but there is a good reason for it, in that we only want to report the findings from those respondents who actually answered the question (and not the seven who did not). That means we have the total of respondents who took part in the survey and a total for those who answered the question. The current example, the difference between the number taking part in the survey and those answering the question is quite small:

35,420 = took part in the survey

35,413 = answered the question 'how long lived in this area'

There are just seven people who did not answer the question, although there can be times when more respondents did not provide an answer than those who did. Let's exaggerate this a little. Let's say we had 1000 people who took part in the survey yet only 1 person provided an answer to one of the questions, it would be important to know this so it can be reported. This is because the data is intended to describe the answers of the sample. In this instance, if only 1 person answered the question, then it would be disingenuous to assume that it represented all who took part in the survey but did not answer the question (the 999). Instances like this can make a huge difference to the percentage found in each possible choice available and leads us to why the IBM SPSS frequency output table provides 'percent' and **'valid percent'**.

Working with valid percent and percent: working with missing data

For the most part, taking part in a survey should be undertaken voluntarily (other than when completing the Census which takes place every 10 years in the UK and is compulsory). The voluntary basis of survey completion leaves us with the systemic problem of missing data. This is when, for whatever reason, respondents have decided not to provide the answer requested of them (how rude!). As stated, this can then mean that there is a huge difference between those who took part in the survey and those who answered a particular question. When making frequency tables in IBM SPSS, such differences are highlighted by providing the valid percent as well as the percent. The difference is in how missing cases (those who did not answer the question) are not included in the valid percent column. Look at Table 4.3 for the variable 'How worried about having car stolen: *wcarstol*'.

Table 4.3 Frequency table for *wcarstol* (How worried about having car stolen)

➡ **Frequencies**

Statistics

How worried about having car stolen

N	Valid	7131
	Missing	28289

How worried about having car stolen

		Frequency	Percent	Valid Percent	Cumulative Percent
Valid	Very worried	430	1.2	6.0	6.0
	Fairly worried	1078	3.0	15.1	21.1
	Not very worried	3703	10.5	51.9	73.1
	Not at all worried	1803	5.1	25.3	98.4
	(Not applicable)	117	.3	1.6	100.0
	Total	7131	20.1	100.0	
Missing	Don't know	1	.0		
	System	28288	79.9		
	Total	28289	79.9		
Total		35420	100.0		

The possible choices range from 'very worried', 'fairly worried', 'not very worried' and 'not at all worried'. There is also the category 'not applicable' to acknowledge that not everyone has a car. Instantly, we recognise this is a categorical ordinal variable.

Table 4.3 is much more complicated and will take a little longer to yield the important information. As before, the first table shows *N Valid 7131* and *Missing 28,289*. This looks different from the *Sex* frequency table (Table 4.1) in that there are a lot of survey respondents who did not answer this question. This becomes more important when the second table is examined and reveals why there are different types of percentages included in the output table. Let's begin to unpack this by looking at the response 'very worried' to the question 'How worried about having car stolen?' The first thing to note is that 430 survey respondents reported being 'very worried' about having their car stolen. If we now look at the next column 'Percent', we see that this equates to 1.2% of the sample (i.e. 1.2% of all those who took part in the survey, i.e. 35,420). However, we need to question just how informative this figure is when so many respondents did not even answer the question. If the next column along is examined, Valid Percent, it shows that the figure has jumped up to 6.0%. The reason why this is higher is because all those people who did not answer the question have been removed. We know this because we have two totals, one where it is all the respondents and one where it is just those who answered the question:

35,420 = all respondents (no 'missing')

7191 = just those who answered the question 'How worried about having car stolen'

Put another way, 6.0% (n = 430) of people who answered the question 'How worried about having car stolen' stated they were 'very worried'. The **valid percent** differs from the percent because it does not include the missing cases.

Reporting the valid percent is more appropriate because the 'missing' cases or all those respondents who did not answer the question were not included in the summary. If the rest of the valid percent data is examined, it shows that 15.15% (n = 1078) report being 'fairly worried', 51.9% (n = 3703) were 'not very worried' and 5.1% (n = 1803) were 'not at all worried'. Include the 'not applicable' of .3% (n = 117) and that adds up to 100%. This is the 100% of those respondents only answering the question and excludes those who did not. It cannot be known why so many people did not answer the question. It might be that they did not because they do not own a car, meaning the question is not applicable. Alternatively, it might be that they had never given it any thought so did not respond. They might have been bored with the questions at this point and just wanted the survey to end. Who knows? While such missing data is interesting, it is a departure from what is important. The main focus has to be on those who answered the question. As such, the valid percent should always be used when reporting the findings from frequency tables.

Box 4.5

Pause for Thought

Why can it be misleading to use the percent rather than the valid percent?

Presenting percentages, don't forget the *n*

The eagle-eyed amongst you will have already noticed that when the valid percent was reported, it was joined by a bracketed number starting with a lower case '*n*'. Why is this needed, we hear you ask. Well it is all about giving as much important information as possible and not trying to mislead whoever is reading the results. Let's take an example about hand washing after using the toilet. If a small-scale study indicated that 75% of teenage boys did not wash their hands after using the toilet, we might think 'yuk' and think twice before shaking any young man's hand. Yet there is other important information that is also required which is the number of people to which the 75% referred. If, for example, it was 3000, then we would rightfully wonder why it was boys did not wash their hands after using the toilet. However, if there were only four respondents, then we might wonder why such findings are being reported: 75% (*n* = 3000) tells a very different story to 75% (*n* = 4). It is therefore important to include this vital piece of information after giving the percent. Both are accurate but would likely lead to very different judgement about the hygiene habits of teenage boys.

What's this cumulative percent column all about?

As well as the percent and valid percent, the final column in the frequency table is called the Cumulative Percent. This can sometimes be confusing but need not be. All the Cumulative Percent column does is to add the percentage of each different value of the valid percent, which should always add up to 100% at the end. Other than that, it has little other use.

To merge or not to merge responses

When presenting findings from frequency tables, as well as reporting the valid percent, it can sometimes be appropriate to merge different categories of responses if it helps develop the narrative. Take another look at the results for the 'How worried car stolen' question. Rather than present the findings from each value, it is sometimes appropriate to merge similar categories in the table summary as shown in Table 4.4.

Table 4.4 Merging data to summarise categories

Old Options	Valid Percent	n	Merged Value Totals	Merged Valid Percent	Merged n
Very worried	6%	430	Worried	21.1%	1508
Fairly worried	15.1%	1078			
Not very worried	51.9%	3703	Not worried	77.2	5506
Not at all worried	25.3%	1803			
(Not applicable)	1.6%	117	N/A		
Total	100%	7131		98.3%	7014

This then makes a more generalised picture of 'worriedness' or 'not worriedness' rather than having four options.

In this case, it would be a simple matter of adding each of the two categories together which would give the following summary of the data:

> When asked how worried the respondents were about having their car broken into, 77.2% (n = 5506) stated that they were 'not very worried' or 'not at all worried', with just over 20% (n = 1508) stating they were either 'very worried' or 'fairly worried'.

The data has not changed other than the 'not applicable' has not been reported in the merged data as clearly, it would not fit in with the overall picture of 'worried-ness'. Whether you decide to merge data in this way is up to you. Generally, it can be done as long as the essence of what is being measured does not change. In some instances, categorical data can be merged, such as with the category of BAME (Black and Minority Ethnic group). In this instance, the rationale for merging these groups could be related to the lower number of such groups throughout society along with the desire to explain if being part of a minority group results in a different societal experience. This is similar to when exploring sexual identity/orientation within the lesbian, gay, bisexual, transgender (LGBT+) communities. Just as there are differences within BAME and LGBT+ communities, there might be occasions when they are grouped together so that a statistical story can be told. A note of caution, however, it is not appropriate to simply merge categorical (nominal) data without providing a strong justification backed up with current literature and data. For instance, it might be easy to think of 'white' and 'non-white' as an appropriate grouping. However, is it appropriate to categorise people by what they are not? This could lead to quite silly groupings. 'Thin' people and 'not-thin' people; 'heterosexuals' and 'non-heterosexuals'; or 'stupid' people and 'non-stupid' people. As such, when or if we merge, it is of the upmost importance that we do so to add to the story. If we run the risk of marginalising an already marginalised group of people, we should think twice and do some research to see what is and is not appropriate. And if in doubt: don't!

Writing up results: producing descriptive summaries

What we have focused on in the chapter thus far is how to work with data. One last task to explore before having a go yourself is how to write up descriptive summaries of the results that have come from the data. When reporting results from frequency tables, it is important to consider how much information is required. Too little and the summary is meaningless; too much and it becomes unwieldy and boring. An important aspect here is to, on the one hand, *state the obvious*, while on the other hand, at least try to tell an interesting story about the data and findings.

There are a few tricks that can be used, although, it should also be noted that with all descriptive summaries, as few words as possible should be used. What we mean here is keep in mind the three Ss (simple, succinct and story). Looking at the 'simple', just because this is statistics doesn't mean it should be complicated. In fact, a real skill of any social researcher is to keep it as simple within the realms of what is being reported. No need to overcomplicate here. Then succinct. Keep it as short as possible. No repetition, no unnecessary words. It might be useful to try and write the descriptive summaries several times and then use the one you think best fits the data. Don't settle for the first one as it might not be the best. Lastly, the story. Keeping in mind simple and succinct, do try and tell a story with what is being reported. Remember, it has to mean something to someone, and the best way to do this is to tell a simple story using the data.

Once you have the three Ss in mind, the next thing to consider when writing descriptive summaries for percentages is to always report the valid percent and then always provide the frequency 'n' so that the reader has enough detail about what the percentage represents. For example,

> 77.2% (n = 5506) stated that they were 'not worried' about having their car stolen.

The above descriptive summary is simple and succinct and has a little story added by merging the data of those who are 'not worried' or 'not at all worried' by the prospect of having their car stolen. The actual percent (77.2%) is given along with the number who has stated this (n = 5506). There are a few other tricks you can use to add a little more interest to the story. Try using the following ideas.

Report the research. As you are expected to write in the third person (meaning you cannot use 'I, me or we' etc.), instead of writing 'my research' (which is first person writing), write 'the research' or 'the data', and so on and then say what the research is (be clear), for example:

> The findings show that 77.2% (n = 5506) stated that they were 'not worried' about having their car stolen (tell the reader what the table is reporting).

Show, compare or present

You can also include a summary statement that will say what the frequency table will do. This can be achieved by following the 'Show, Compare or Present' rules (mix and match these ideas with various tables – don't use them all for one table):

> 'The results from the question asking about how worried respondents were about having their car stolen show . . .'

'When comparing the results of those who were and were not worried about having their car stolen . . .'

'Table 4.3 presents the respondents' perception about how worried they were about having their car stolen . . .'

Mix and match these and see which works the best. From here, you can highlight any areas that are interesting or data that is surprising, for example:

'The findings in Table 4.3 show that while 21.1% (n = 1508) stated they were "very or fairly worried" about having their car stolen, 77.2% (n = 5506) stated they were not'.

Or

'What is interesting about the data in Table 4.3 is how over 77% (n = 5506) of respondents were "not worried" about having their car stolen'.

Or

'The most striking results to emerge from the data presented in Table 4.3 is the high numbers of people who were "not worried" about having their car stolen (72.2%, n = 5506) and, as such, challenge the current perception of fear of car crime . . .'

Descriptive summaries can be even harder than the statistics themselves because this is where you have to do the work; statistical software cannot write it for you. But it is worth the effort.

Rounding up or down?

There are no hard and fast rules about rounding the results up or down. If you look at the valid percent reported above, we might ask ourselves if 77.2% and 21.1% need rounding down to 70% and 21%. This is a decision you have to take. It is worth checking this with your tutor before doing this. As a rule of thumb, if you are to round up or down, it is down if the figure is below .49 (e.g. 7.49 or below could be rounded down to 7) whereas .51 can be rounded up (e.g. 7.5 and above could be rounded up to 8). However, looking at the example of 70.1% and 21.1%, there seems little point in rounding up or down as the figures are accurate and not too unwieldly. The best rule might be 'if in doubt; don't!'

What to report with missing data?

With some survey data, it might be tempting to state the number of 'missing'. For instance, a question on a sensitive topic (e.g. sexual orientation) might result in many people not answering the question. It is extremely interesting when this happens: But is this a finding? Possibly. However, it is better to report data from those who actually provided an answer rather than focus on those who do not, as we cannot know why they chose to not answer the question. It is therefore not necessary to report the missing data in the descriptive summary.

Box 4.6

Time to Get Your Hands Dirty! Interpreting Frequency Tables

It's now your turn to have a go at interpreting the data in a frequency table. Table 4.5 comes from the CSEW 2016–2017. The question asked 'How worried about being physically attacked by strangers'. Remember to take a look at how many missing cases there are but not to get distracted by them. Have a go at presenting the findings from the table as illustrated above. Just tell the story of what you see. Remember, all you need to do is 'state the obvious' and use the three Ss (simple, succinct and story) to report what you think is the most important result. Often, this could be the finding with the highest valid percent. When you are ready, open up an MS Word document and make a table similar to Table 4.4 that has merged the data about worries over car being stolen. Use the findings in the frequency table (Table 4.5). It is good practice to put the old values in so that the calculations can be checked at a later stage.

Table 4.5 How worried about being physically attacked

⇒ **Frequencies**

Statistics

How worried about being physically attacked by strangers

N	Valid	8748
	Missing	26672

How worried about being physically attacked by strangers

		Frequency	Percent	Valid Percent	Cumulative Percent
Valid	Very worried	688	1.9	7.9	7.9
	Fairly worried	1431	4.0	16.4	24.2
	Not very worried	4163	11.8	47.6	71.8
	Not at all worried	2439	6.9	27.9	99.7
	(Not applicable)	27	.1	.3	100.0
	Total	8748	24.7	100.0	
Missing	Refused	1	.0		
	Don't know	4	.0		
	System	26667	75.3		
	Total	26672	75.3		
Total		35420	100.0		

Strengths of using the percent

We hope by now you can see the value of descriptive statistics, frequency tables and in this instance, percentages. Percentages when reported correctly are extremely powerful. As will be discussed in Chapter 7, it is by using simple statistics correctly that social science students are able to develop and tell a convincing story that uses quantitative evidence. Of course, there are other reasons why percentages are useful. One of the key strengths is how it is possible to compare groups/samples or populations of different sizes. We can do this because percentages are a standardised approach to reporting data. To illustrate this point, take a look at Figure 4.7 that reports the number of recorded homicides in England and Wales where a knife or sharp instrument was used between the years 2012 and 2019 (up to April 2019 which, in this instance, is the full year as this data set uses the tax year as its start and end point).

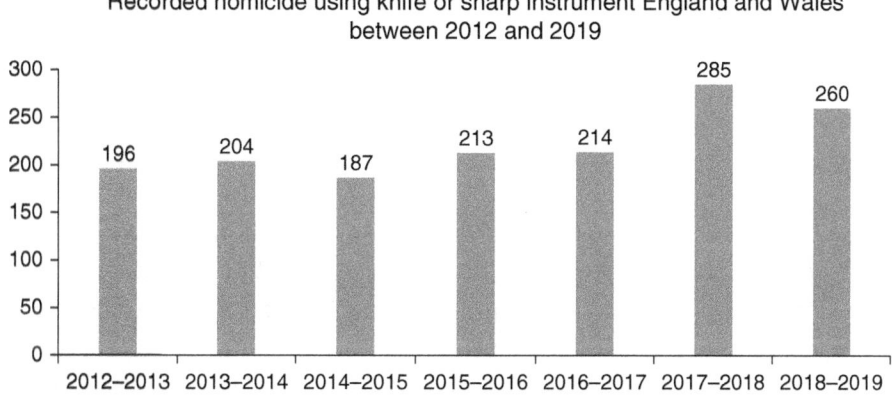

Figure 4.7   Year on year comparisons of recorded homicide using knife or sharp instrument - frequencies

Source. Data adapted from Police Recorded Crime-knife-open-data-march2009-onwards-tables

Looking at Figure 4.7, what are the issues that stand out the most? Is it that in each year, the number of recorded homicides is different? In 2012–2013 it was 196 homicides. It went up to 204 in 2013–2014 and back down to 187 in 2014–2015. Can you see that year on year, there does appear to be an upward trend with the number of homicides peaking in 2017–2018? However, the number of homicides actually went down the year after in 2018–2019. Can we assume that this type of crime is starting to decline? As a social science student, your job is to scrutinise this data and ensure we are telling a complete story and not using selective data. As you might imagine, the above graph is telling an incomplete story. Can you think of what other data is needed to be able to make a better comparison?

For one thing, you might also want to know the number of other crimes that are committed using a knife or sharp instrument. Figure 4.8 shows *homicide with injury and assault with intent to cause serious harm.*

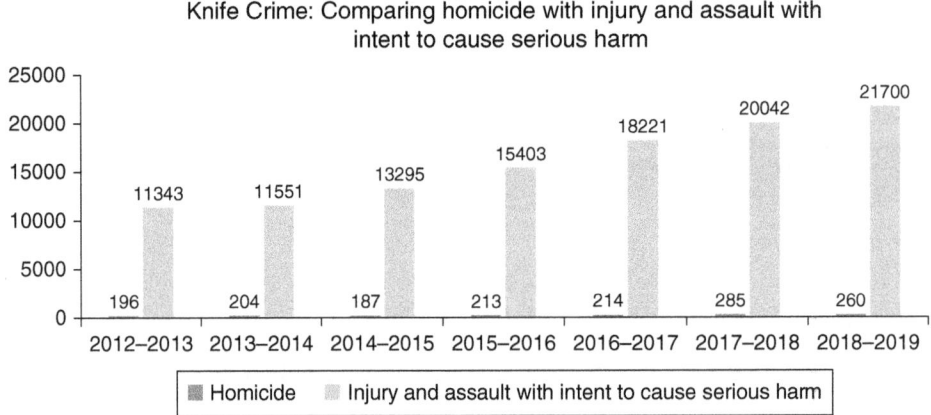

Figure 4.8 Year on year comparisons of recorded homicide using knife or sharp instrument versus assault – frequencies

Source. Data adapted from Police Recorded Crime-knife-open-data-march2009-onwards-tables

Now we have much more information to work with from which to make our judgements. Looking at Figure 4.8 shows that there is a clear upward trend of crimes whereby knifes or sharp instruments are used at least for these two crimes. However, when making comparisons using 'raw' data, it is not possible to make appropriate comparisons, and this is where percentages are most useful. Figure 4.9 expresses the data as percentages.

Whereas Figure 4.8 made it seem that the fewest number of homicides were recorded in 2014–2015, when we see this expressed as a percentage (Figure 4.9), it shows that it was actually the same as 2015–2016 and, in fact, higher than 2016–2017 and 2018–2019. Remember, one of the key strengths of using percentages is how groups/samples or populations of different sizes can be compared. By sharing this common baseline of percentages then allows for the proper comparisons between groups (in this instance, years) to be made. The main issue here is that when dealing with findings from data sets, it is important to remember that the story being told provides as much detail as possible for the reader to make their own judgement. In essence, the findings should speak for themselves, meaning you must give them a voice to do so in as honest a way as possible.

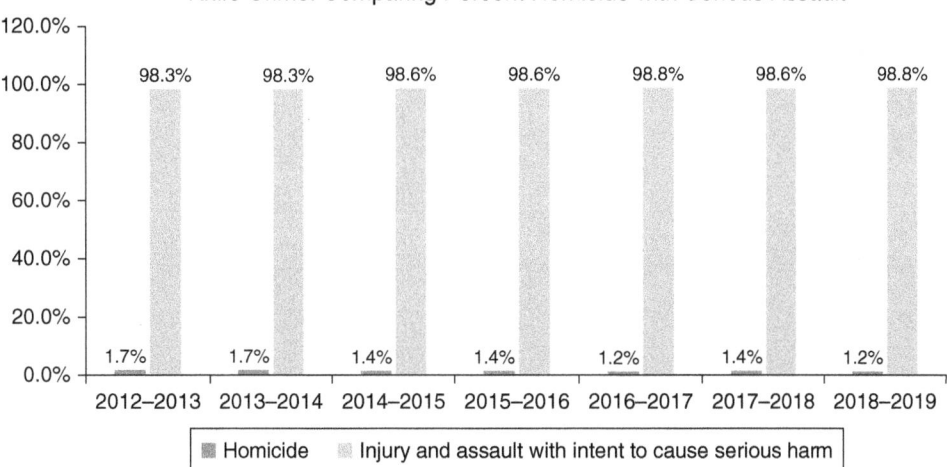

Figure 4.9 Year on year comparisons of recorded homicide using knife or sharp instrument versus assault – percentages

Source. Data adapted from Police Recorded Crime-knife-open-data-march2009-onwards-tables

Working with contingency tables

Thus far, we have explored how to use just one variable, hence the name univariate analysis. But data can get really exciting when two variables are used together to tell a story. This is called bivariate analysis (the prefix 'bi' is derived from Latin meaning 'two'). When using two variables in a table like this, it is called either a 'crosstab table', 'cross-tabulation' or 'contingency table'. They are all the same thing with different names, just make sure you consistently use the same one, whichever one you choose. So, while univariate analysis does a great job at allowing us to explore a single variable, it does have limitations in that the story we can ultimately tell might have important information missing. For instance, look back at Table 4.3 where you explored how worried respondents were about having their car stolen. Could we be missing something important here? Think about it. Are we saying that all people, men/women old/young feel the same way? We might think that other important factors such as gender or age group might also play an important part in the degree of worry about having their car stolen. Using univariate analysis only allows us to look at individual variables, but wouldn't it be amazing if we could put them together to see if men and women or older and younger people feel differently? This is what is meant by making crosstab tables so that bivariate analysis can be carried out on more than one variable. The strength of crosstab tables comes from being able to compare findings from different subgroups. The output Table 4.6 is the crosstab table for 'How worried about having car stolen' combined with sex, the subgroup of men and women to reveal if there are any differences by gender.

Table 4.6 IBM SPSS output table for 'How worried about having car stolen' by sex

How worried about having car stolen * Adult number 1 (respondent): Sex Crosstabulation

Count

		Adult number 1 (respondent): Sex		
		Male	Female	Total
How worried about having car stolen	Very worried	167	263	430
	Fairly worried	482	596	1078
	Not very worried	1742	1961	3703
	Not at all worried	900	903	1803
	(Not applicable)	57	60	117
Total		3348	3783	7131

With Table 4.6, the results are broken down into the subgroups of men and women. This means we can start to compare if the experience is the same for men and women. And look, female respondents consistently feel more worried about having their car stolen when compared to men. They also feel less worried which is confusing. Of course, there are more women in this sample (3783 women compared to 3348 men) so this might account for the differences. Can you think of how this data can be better explored? Think back to how 'raw' data can sometimes be misleading or misinterpreted, and what is actually needed to make a fairer comparison is the valid percent. This is shown in Table 4.7.

Table 4.7 IBM SPSS output table for 'How worried about having car stolen' by sex (including column percent)

How worried about having car stolen * Adult number 1 (respondent): Sex Crosstabulation

			Adult number 1 (respondent): Sex		
			Male	Female	Total
How worried about having car stolen	Very worried	Count	167	263	430
		% within Adult number 1 (respondent): Sex	5.0%	7.0%	6.0%
	Fairly worried	Count	482	596	1078
		% within Adult number 1 (respondent): Sex	14.4%	15.8%	15.1%
	Not very worried	Count	1742	1961	3703
		% within Adult number 1 (respondent): Sex	52.0%	51.8%	51.9%
	Not at all worried	Count	900	903	1803
		% within Adult number 1 (respondent): Sex	26.9%	23.9%	25.3%
	(Not applicable)	Count	57	60	117
		% within Adult number 1 (respondent): Sex	1.7%	1.6%	1.6%
Total		Count	3348	3783	7131
		% within Adult number 1 (respondent): Sex	100.0%	100.0%	100.0%

By including the column percent allows us to more thoroughly explore the differences between men and women and their worries over having their car stolen. It is not shown that women are more worried with 7% (n = 263) being very worried compared to 5% (n = 167) of men. We can go even further. Look at the difference between men and women and whether they are 'not at all worried'. Again, we can see that there are 26.9% (n = 900) men compared to 23.9% (n = 903) women. Can you see that combining all the skills above has allowed you to develop the story about how worried respondents were about having their car stolen. Not only can we see that there are more women than men who are worried, by using the percent, this is confirmed and accounts for the difference in numbers between men and women.

How to guides for IBM SPSS and MS Excel

Before we begin, imagine you meet and fall in love with someone from another country where you did not speak the same language. The chances are, you will be able to communicate at some level (the language of love?), but things might be a little difficult while you learn this new way to communicate. Well, in a similar way, learning how to use a new statistical package such as IBM SPSS or MS Excel is just the same as learning a new language. It takes practice and patience. At first, things seem very strange and it would be easy to give up. Don't! Give it time and soon you'll be whizzing around each package like a pro.

Before we start exploring each package, be aware that IBM SPSS and MS Excel are very different beasts when it comes to carrying out analysis and producing frequency tables. It is for this reason, that we have kept the 'How to' guide separate. The first thing to note is how MS Excel is a Spreadsheet package that can carry out statistical analysis whereas IBM SPSS is an industry standard statistical package. That is not to say that one is better than the other: they are just different in how to get them to do the analysis on the data set. When using MS Excel, there are a variety of ways to make frequency tables, although we think the Pivot Table function offers the quickest and easiest way. IBM SPSS has a specific 'Analyse' function that allows for quick and easy statistical analysis with very little knowledge needed. In fact, it is a simple matter of 'pointing and clicking' as will be demonstrated now. To do this, we will use the same variables as we used above starting with IBM SPSS.

How to guide for IBM SPSS

The first thing to do is to access and download the CSEW 2016–2017 in IBM SPSS – Chapter 2 outlines how to access data sets and you can find the CSEW 2016–2017 on the UKDS website (www.ukdataservice.ac.uk). Assuming you have the software package on your computer, then clicking on the link will launch IBM SPSS. We are using a version we cleaned to make it easier for us to use. Think back to Chapter 2 and the principles set out there on how to manage large data sets once you have downloaded this file.

Once you have IBM SPSS open, follow the below instructions to carry out univariate analysis on the variable '*hrprel3* – HRP religion'.

Carrying out univariate analysis

First go to **Analyze** (American spelling on IBM SPSS) on the toolbar and click **Descriptive Statistics – Frequencies** from the drop-down menu, as shown in Figure 4.10.

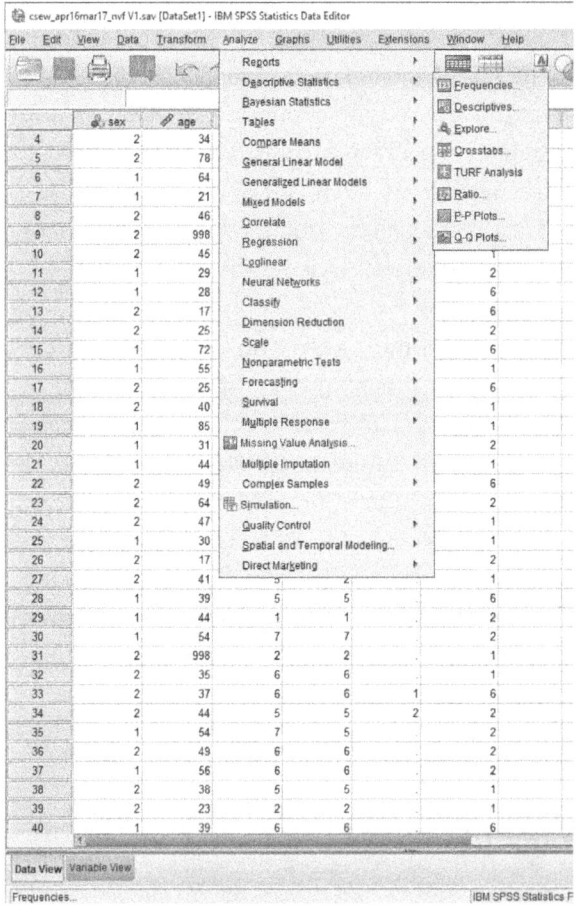

Figure 4.10 Using Analyze on the IBM SPSS toolbar to access frequencies

The 'Frequencies' screen shown in Figure 4.11 appears.

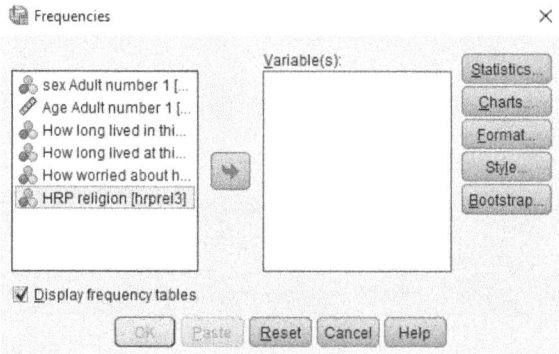

Figure 4.11   The Frequencies dialog box in IBM SPSS

In the dialog box, drag *HRP religion [hrprel3]* from the left-hand window to the right under *Variable(s)* and then Click OK (as shown in Figure 4.12).

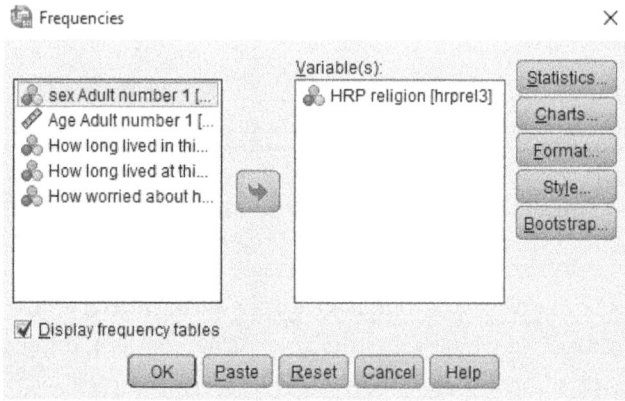

Figure 4.12   Selecting a variable in the Frequencies dialog box in IBM SPSS

Then a separate output table will appear similar to the ones shown earlier in the chapter.

IBM SPSS outputs

Using the techniques shown above, you should now have frequency table as shown in Table 4.8; work through both tables familiarise yourself with the outputs. By using the same data as above, you can check that your outputs look the same. Remember to write the descriptive summary of what you think is important in the analysis.

Table 4.8 IBM SPSS outputs for HRP religion (CSEW 2016–2017)

➡ **Frequencies**

Statistics

HRP religion

N	Valid	35414
	Missing	6

HRP religion

		Frequency	Percent	Valid Percent	Cumulative Percent
Valid	No religion	12360	34.9	34.9	34.9
	Christian	20544	58.0	58.0	92.9
	Buddhist	170	.5	.5	93.4
	Hindu	440	1.2	1.2	94.6
	Jewish	145	.4	.4	95.0
	Muslim	1232	3.5	3.5	98.5
	Sikh	199	.6	.6	99.1
	Other	180	.5	.5	99.6
	Refused	86	.2	.2	99.8
	Don't know	58	.2	.2	100.0
	Total	35414	100.0	100.0	
Missing	System	6	.0		
Total		35420	100.0		

Note. CSEW = Crime Survey of England and Wales.

Once you have done this, have another go with different variables. You could use the variable 'Sex' and then 'How long lived at this address'. Remember that this chapter focuses on categorical variables rather than scale. As such, don't do any analysis on the scale variable 'Age'.

Two's company: carrying out bivariate analysis using IBM SPSS

When using IBM SPSS, the crosstab table is made automatically (unlike when using MS Excel where there is a very different process to making them). Let's do one building on what has been done thus far.

In the above univariate analysis, the variable 'religion' was explored with a frequency table showing the degree to which different religions, or none, were followed. While this is informative, we can go further. Let's say we also wanted to know if there was a difference between men and women in terms of their levels of religious identity. All that is needed is for both variables to be used to make a crosstab table. It is these crosstab tables that allow us to compare different subgroups, in this instance, compare 'men' and 'women' and what religion they follow. Once you have carried out the univariate analysis using IBM SPSS, it is not that difficult to move up to bivariate analysis. The first part of the procedure is the same although there are a few extra steps that are required. First, go to **Analyze** on the toolbar and click **Descriptive Statistics – Crosstabs** from the drop-down menu, as shown in Figure 4.13.

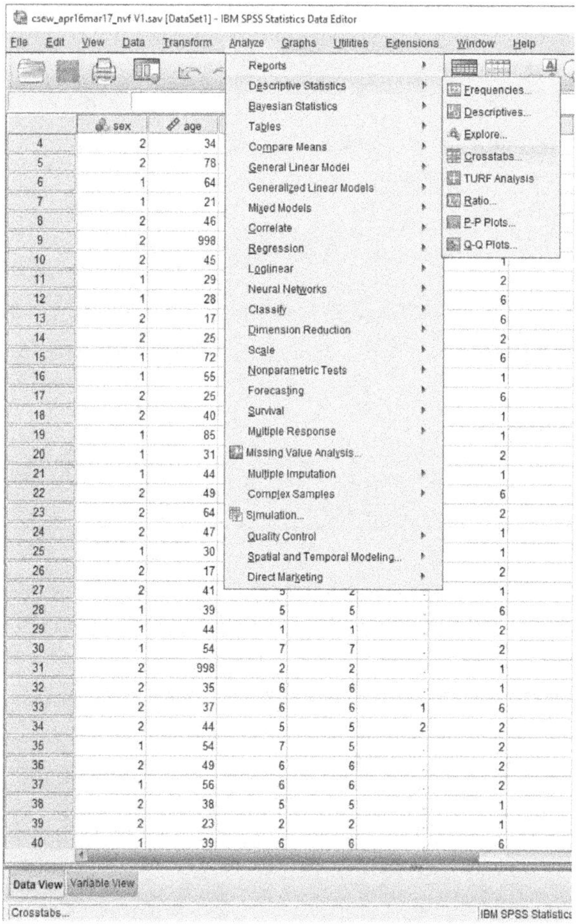

Figure 4.13   Using Analyze on the IBM SPSS toolbar to access crosstabs

Then in the new dialog box, drag *HRP religion [hrprel3]* from the left-hand window to the right-hand side and place in **Rows**. Then drag Sex from the left-hand window to the right-hand side and place in **Column** as shown in Figure 4.14. Then click **OK** and a separate output table (Figure 4.15) will appear similar to the ones shown earlier in the chapter.

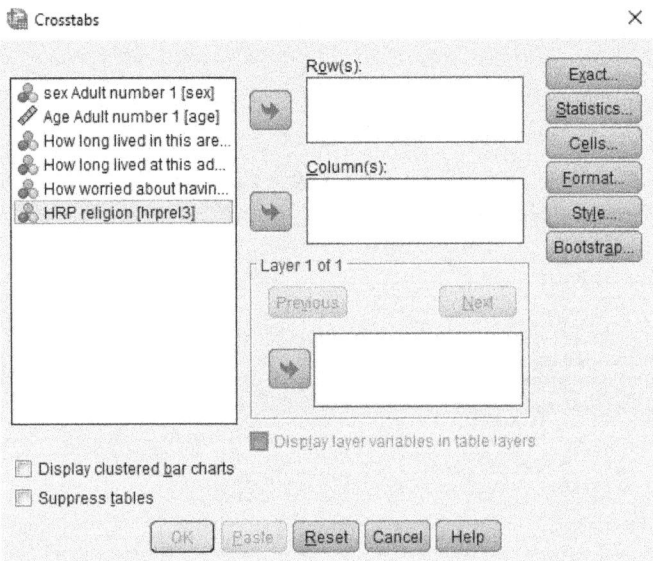

Figure 4.14   The Crosstabs dialog box in IBM SPSS

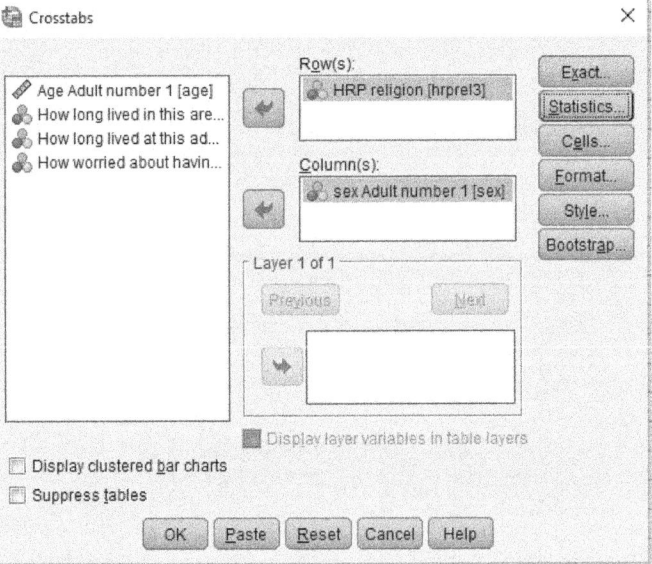

Figure 4.15   Selecting variables in the Crosstabs dialog box in IBM SPSS

A quick note on **Rows** and **Columns** and which variable goes where. It does matter which variable you put in the row and column. The column is where you put the independent variable (IV) and the row is for the dependent variable (DV). In the above instance, a person's religion (DV) is dependent on their sex (male/female) (IV).

One way to check that the IV and DV have been put in the correct place is to see which is at the top of the table and which is at the side. If the IV is at the top, then this is correct. If the IV is at the side, then this is wrong.

As can be seen in Table 4.9, it has two different variables and has *Sex* at the top of the table. Take some time and explore all the data and findings. Once you have done the above bivariate analysis, have another go with a different variable.

Table 4.9 Crosstab for religion and sex

⇨ **Crosstabs**

Case Processing Summary

	Cases					
	Valid		Missing		Total	
	N	Percent	N	Percent	N	Percent
HRP religion * sex Adult number 1	35414	100.0%	6	0.0%	35420	100.0%

HRP religion * sex Adult number 1 Crosstabulation

Count

		sex Adult number 1		
		Male	Female	Total
HRP religion	No religion	6164	6196	12360
	Christian	9019	11525	20544
	Buddhist	76	94	170
	Hindu	247	193	440
	Jewish	71	74	145
	Muslim	623	609	1232
	Sikh	95	104	199
	Other	73	107	180
	Refused	40	46	86
	Don't know	26	32	58
Total		16434	18980	35414

Bivariate analysis: including percentages

The eagle-eyed amongst you might have noticed that the crosstab table (Table 4.9) only includes the 'raw' data and does not include the percentage. If we wanted to compare men and women's religious denomination, then it would be more appropriate to include the percent. To do this in IBM SPSS, you follow the exact same process as you did for running a crosstab, except this time you add one step, which is to click on **Cells** on the right hand side of the Crosstabs dialog box. This opens a new dialog box as shown in Figure 4.16.

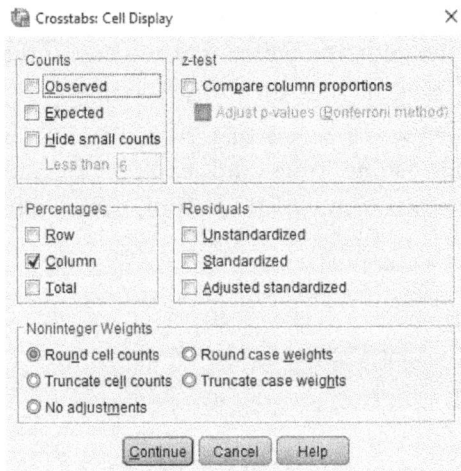

Figure 4.16 The Crosstabs: Cell Display dialog box in IBM SPSS

Under Counts, unclick 'Observed' and Under **Percentages**, click on 'Column'; then click OK to run the table (Figure 4.17).

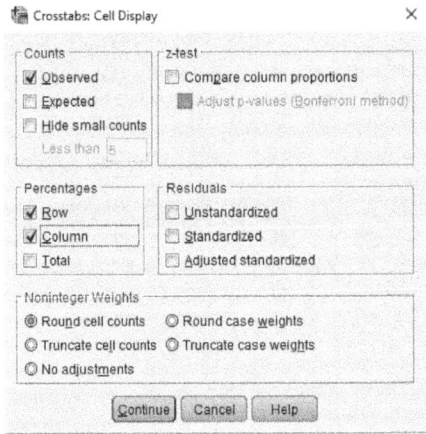

Figure 4.17 Selecting percentages and counts in the Crosstabs (Cell Display) dialog box in IBM SPSS

Table 4.9 is a crosstab for religion and sex, whereas Table 4.10 includes column and row percent.

In IBM SPSS, it is possible and very easy to show the Column percent, the Row percent as well as the 'raw' data (the Observed count) all in one table. It is a matter of ticking the appropriate box as illustrated in Figure 4.17.

Table 4.10 Crosstab for religion and sex including percentages

Case Processing Summary

	Cases					
	Valid		Missing		Total	
	N	Percent	N	Percent	N	Percent
HRP religion * Adult number 1 (respondent): Sex	35414	100.0%	6	0.0%	35420	100.0%

HRP religion * Adult number 1 (respondent): Sex Crosstabulation

Count

		Adult number 1 (respondent): Sex		Total
		Male	Female	
HRP religion	No religion	6164	6196	12360
	Christian	9019	11525	20544
	Buddhist	76	94	170
	Hindu	247	193	440
	Jewish	71	74	145
	Muslim	623	609	1232
	Sikh	95	104	199
	Other	73	107	180
	Refused	40	46	86
	Don't know	26	32	58
Total		16434	18980	35414

The crosstab we have now run (Table 4.11) with the Column, Row and Count data and contains such a lot of data that it might be difficult to work through and notice the key information. It is a matter of choice as to how much to include in a table. However, it has to be noted that when presenting findings to others, it is good practice to keep the table as simple as possible; a point which we will return to in Chapter 6. Here we strongly suggest that less is more.

How to guide for MS excel

Now that you have become familiar with IBM SPSS, let's now look at how the same analysis can be undertaken using MS Excel. MS Excel is part of the larger MS package, which is the most commonly used software globally. We also noted that a lot of the data we might want to use may be in MS Excel format, so it is good to know some MS Excel basics. As noted earlier, MS Excel is quite different in that it is not a

Table 4.11 Crosstab for religion and sex showing column percentages

HRP religion * Adult number 1 (respondent): Sex Crosstabulation

			Adult number 1 (respondent): Sex		
			Male	Female	Total
HRP religion	No religion	Count	6164	6196	12360
		% within HRP religion	49.9%	50.1%	100.0%
		% within Adult number 1 (respondent): Sex	37.5%	32.6%	34.9%
	Christian	Count	9019	11525	20544
		% within HRP religion	43.9%	56.1%	100.0%
		% within Adult number 1 (respondent): Sex	54.9%	60.7%	58.0%
	Buddhist	Count	76	94	170
		% within HRP religion	44.7%	55.3%	100.0%
		% within Adult number 1 (respondent): Sex	0.5%	0.5%	0.5%
	Hindu	Count	247	193	440
		% within HRP religion	56.1%	43.9%	100.0%
		% within Adult number 1 (respondent): Sex	1.5%	1.0%	1.2%
	Jewish	Count	71	74	145
		% within HRP religion	49.0%	51.0%	100.0%
		% within Adult number 1 (respondent): Sex	0.4%	0.4%	0.4%
	Muslim	Count	623	609	1232
		% within HRP religion	50.6%	49.4%	100.0%
		% within Adult number 1 (respondent): Sex	3.8%	3.2%	3.5%
	Sikh	Count	95	104	199
		% within HRP religion	47.7%	52.3%	100.0%
		% within Adult number 1 (respondent): Sex	0.6%	0.5%	0.6%
	Other	Count	73	107	180
		% within HRP religion	40.6%	59.4%	100.0%
		% within Adult number 1 (respondent): Sex	0.4%	0.6%	0.5%
	Refused	Count	40	46	86
		% within HRP religion	46.5%	53.5%	100.0%
		% within Adult number 1 (respondent): Sex	0.2%	0.2%	0.2%
	Don't know	Count	26	32	58
		% within HRP religion	44.8%	55.2%	100.0%
		% within Adult number 1 (respondent): Sex	0.2%	0.2%	0.2%
Total		Count	16434	18980	35414
		% within HRP religion	46.4%	53.6%	100.0%
		% within Adult number 1 (respondent): Sex	100.0%	100.0%	100.0%

'point and click' process. The main feature we suggest when using MS Excel for categorical data are Pivot Tables. To get to Pivot Tables, however, requires a little more upfront work and, in some instance, a little knowledge about the formula (more so in the next chapter when using scale data). Don't let this put you off as formula can be as simple as '=A2 + A3 + A4' in MS Excel. Another key difference about MS Excel is how it can work with both numerical and text data, unlike IBM SPSS, which can only work with numbers. For instance:

1   Men
2   Women

As MS Excel does not distinguish between the different types of data, it can use either numeric or text depending on what statistics are needed. When using categorical data in MS Excel, it is much easier to use the actual text (e.g. male or female rather than 1 or 2) which for a novice researcher can be easier and more intuitive. The text name is then used in the subsequent Pivot Table function. One of the main advantages of using Pivot Tables is how the data can still be manipulated with cases such as 'don't know' or 'refusal' being easily removed.

Using MS word to make frequency tables

Let's start by using our fictional data (*Ice cream dataset 2019*) to illustrate how the data looks in MS Excel.

Figures 4.18 and 4.19 are the equivalent screenshots of the data set in MS Excel showing the top and bottom responses, respectively.

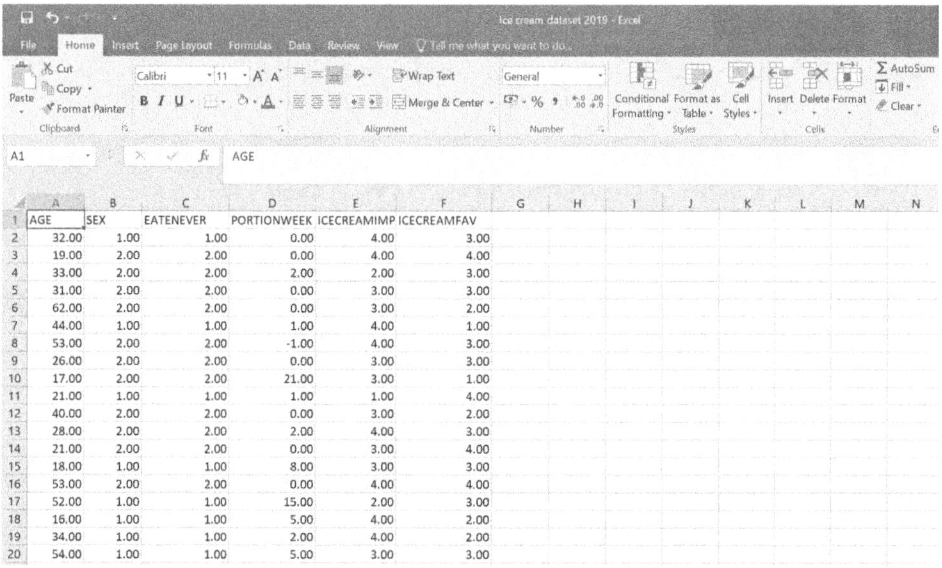

	A	B	C	D	E	F
1	AGE	SEX	EATENEVER	PORTIONWEEK	ICECREAMIMP	ICECREAMFAV
2	32.00	1.00	1.00	0.00	4.00	3.00
3	19.00	2.00	2.00	0.00	4.00	4.00
4	33.00	2.00	2.00	2.00	2.00	3.00
5	31.00	2.00	2.00	0.00	3.00	3.00
6	62.00	2.00	2.00	0.00	3.00	2.00
7	44.00	1.00	1.00	1.00	4.00	1.00
8	53.00	2.00	2.00	-1.00	4.00	3.00
9	26.00	2.00	2.00	0.00	3.00	3.00
10	17.00	2.00	2.00	21.00	3.00	1.00
11	21.00	1.00	1.00	1.00	1.00	4.00
12	40.00	2.00	2.00	0.00	3.00	2.00
13	28.00	2.00	2.00	2.00	4.00	3.00
14	21.00	2.00	2.00	0.00	3.00	4.00
15	18.00	1.00	1.00	8.00	3.00	3.00
16	53.00	2.00	2.00	0.00	4.00	4.00
17	52.00	1.00	1.00	15.00	2.00	3.00
18	16.00	1.00	1.00	5.00	4.00	2.00
19	34.00	1.00	1.00	2.00	4.00	2.00
20	54.00	1.00	1.00	5.00	3.00	3.00

Figure 4.18 Ice cream data set 2019 in MS Excel (top view)

In order to explore how to run frequency tables in MS Excel, we will use some real data (CSEW 2016–2017); this is the data we used for our IBM SPSS examples earlier in this chapter and we will use the same six variables that have been imported into MS Excel. You can download this data in MS Excel format from the UKDS website, see Chapter 2 for more information. One of the easiest ways to make a frequency table in MS Excel is to turn the data into a Pivot Table. In the later versions of MS Excel (2010 onward), this is a simple matter of selecting which data is to be used, clicking on 'Insert', and selecting 'Pivot Table'. It should be noted that for this to be successful, each column (the variable) must have a name (it cannot be left blank).

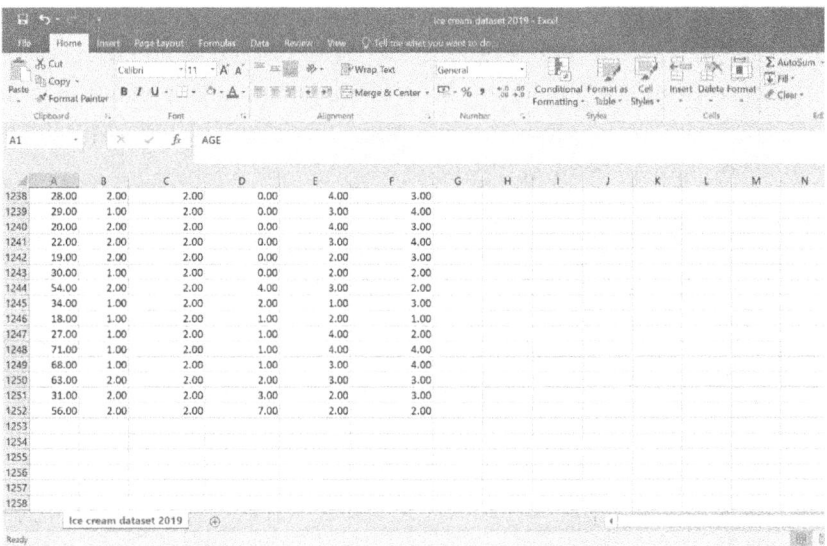

Figure 4.19 Ice cream data set 2019 in MS Excel (bottom view)

Follow the steps below:

Step 1: Select data (see Figure 4.20) to be used in Pivot Table – *Sex*.

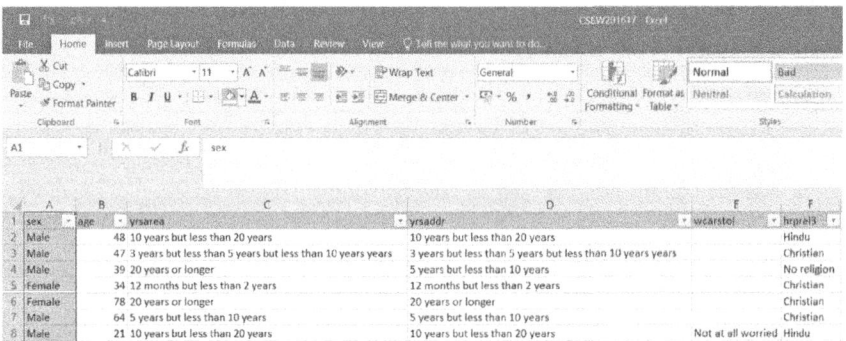

Figure 4.20 Selecting a variable in MS Excel

Step 2: Click on the 'Insert' tab, as shown in Figure 4.21.

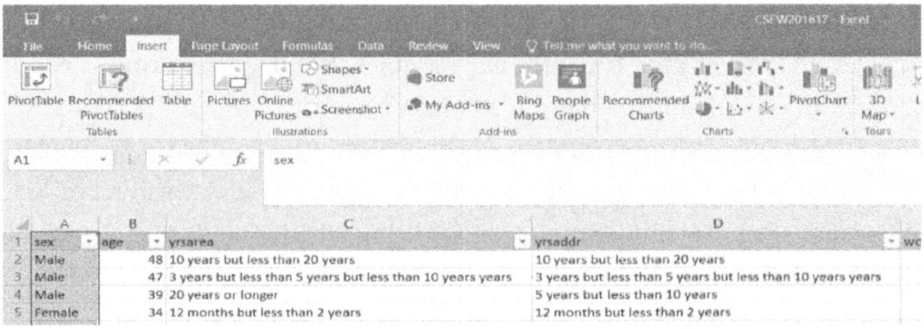

Figure 4.21 Using the Insert Tab in MS Excel

Step 3: Click PivotTable. This will bring the following dialog box up 'Create PivotTable', as shown in Figure 4.22.

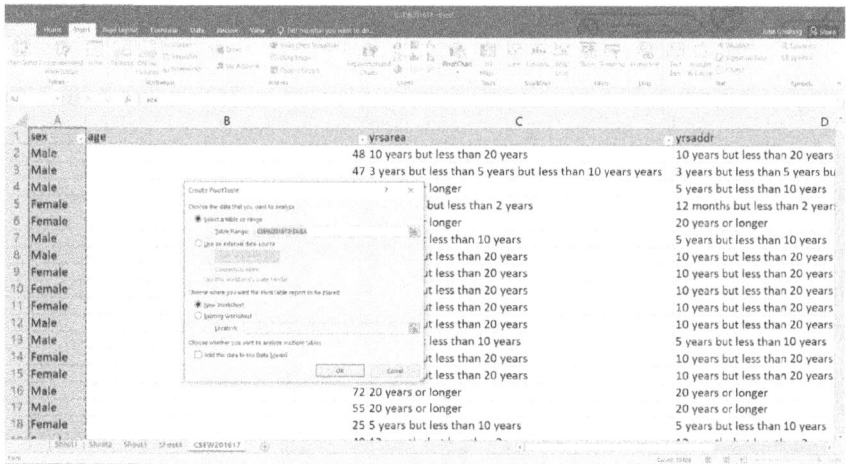

Figure 4.22   The Create Pivot Table dialog box in MS Excel

Step 4: Check that the 'New Worksheet' is selected.

Step 5: Click OK. This will take you to a new Worksheet (Figure 4.23) where you can make the PivotTable.

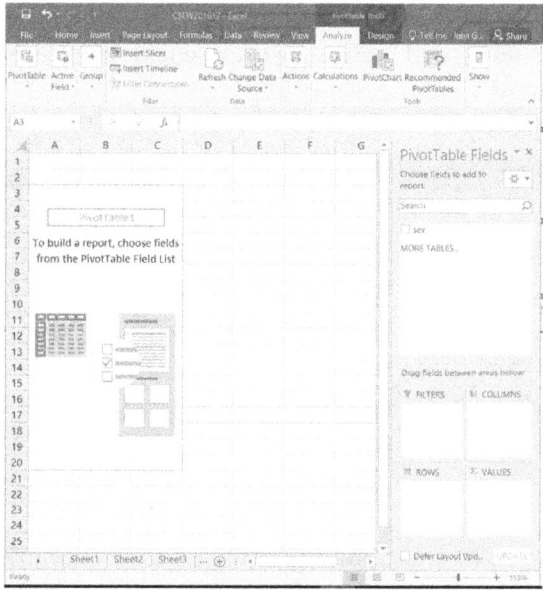

Figure 4.23   Blank Worksheet for a Pivot Table in MS Excel

The new worksheet will have the additional PivotTable Fields. As only one variable has been selected, only '*Sex*' is shown.

Step 6: Drag and drop Sex into both 'Rows' and 'Values' as shown in Figure 4.24.

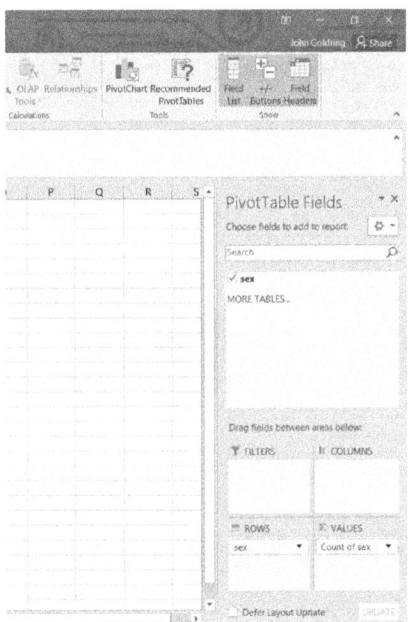

Figure 4.24 The Pivot Table Fields dialog box in MS Excel

Notice that in 'Rows', the variable *'sex'* has been placed while in 'Values', it is the 'Count of sex'. Both are needed to produce the Pivot Table. Once the variable has been placed in the 'Rows' and 'Values', it will then allow MS Excel to automatically make the PivotTable as shown in Figure 4.25.

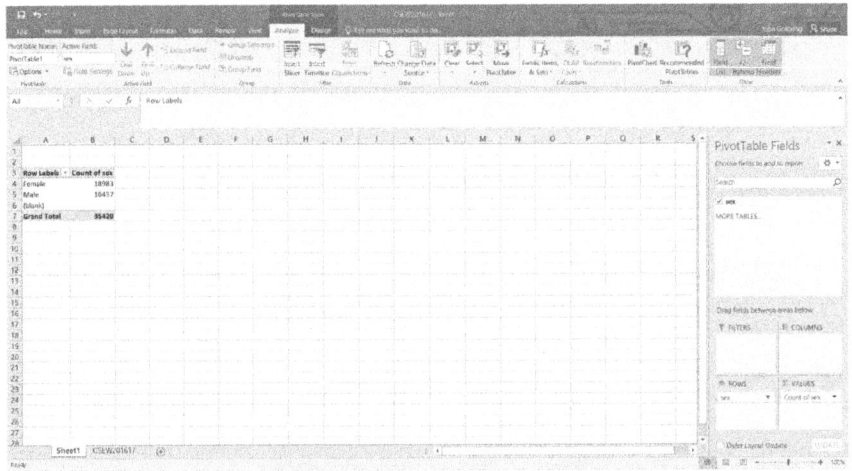

Figure 4.25 Pivot Table of sex (CSEW 2016–2017) in MS Excel

Note. CSEW = Crime Survey of England and Wales.

In this example, there are no missing data. It shows that there are 18,983 females and 16,437 males. The ('blank') and where necessary any missing data should be filtered out by clicking on the small triangle next to the Row Labels. Doing this will result in just the different categories being shown, as Figure 4.26 demonstrates.

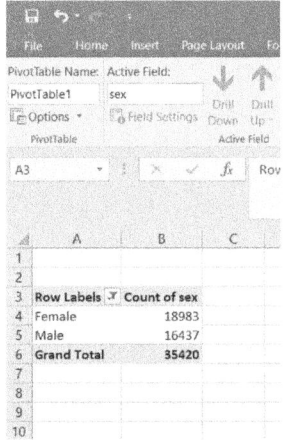

Figure 4.26 Pivot Table of sex (CSEW 2016–2017) in MS Excel, with (blank) removed

Note. CSEW = Crime Survey of England and Wales.

A key difference between IMB SPSS and MS Excel is the amount of detail that is shown in a frequency table compared to a Pivot Table. Just by looking at the Pivot Table for *sex* above reveals that only the count of *sex* is shown. As discussed earlier, this is not enough and what is needed is the Valid Percent. As luck (or design) would have it, accessing percentages in the PivotTable is not difficult and only requires a few extra clicks. However, doing so will essentially change your count into the percent meaning both will not be shown. A simple solution to this is to copy and paste the count before changing them. See the steps below:

Step 1: Copy and paste the Count of sex into column C (Figure 4.27).

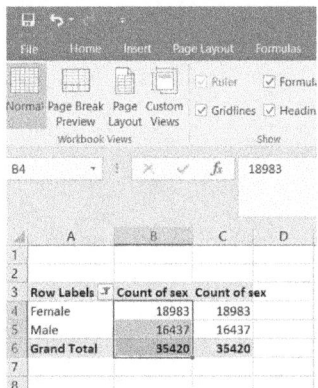

Figure 4.27 Creating a Copy of the Variable Count in MS Excel

Step 2: Reselect the data in Column B (Figure 4.28).

Step 3: Right click the mouse to bring up the following window (Figure 4.28).

Step 4: Select 'Show Values As' and then '% of Grand Total'. This will change the count into the percentage (Figure 4.28).

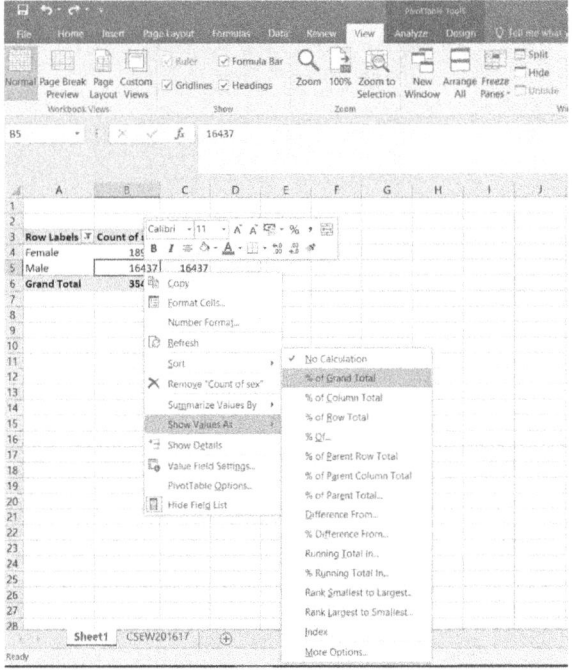

Figure 4.28 Converting counts into percentages in MS Excel

The PivotTable (Table 4.12) will now show the valid percent for males and females assuming that the blanks and missing cells have been filtered out.

Table 4.12 Count of sex showing valid percent

	A	B	C	D
1				
2				
3	Row Labels	Count of sex	Count of sex	
4	Female	53.59%	18983	
5	Male	46.41%	16437	
6	Grand Total	100.00%	35420	
7				
8				
9				
10				
11				
12				

Once you have done this, write up the descriptive summary of the findings. Why not compare the Pivot Table with the Frequency Table from IBM SPSS, which do you prefer? Now have a try with a variable that has more values. Remember, frequency tables using percentages are for categorical data, so do not use the scale 'age'. The next example will use the variable 'hrprel3' that measures 'religion'. Following the same steps as above should lead you to create a PivotTable similar to Figure 4.29. Remember, select the variable you want or all of them before going to 'Insert' and 'PivotTable'.

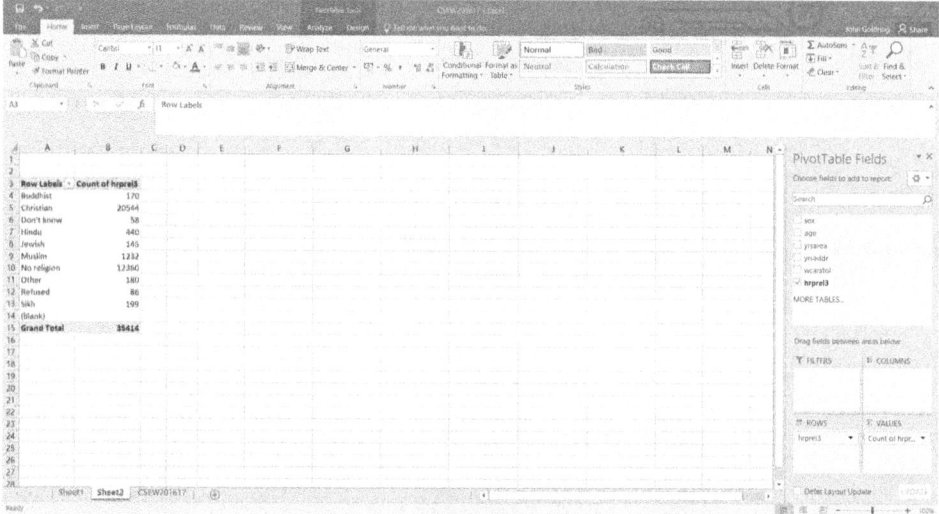

Figure 4.29   Pivot Table of *hrprel3* (CSEW 2016–2017) in MS Excel

Note. CSEW = Crime Survey of England and Wales.

In the above example, all the variables were selected but only 'hrprel3' was dragged and dropped into the 'Rows' and 'Values'. If you look carefully at the table, you will notice that 'Don't know', 'Other', 'Refused' and '(blank)' are showing. These need to be filtered out in the same way that the blank were filtered out of 'sex' above. Once done, the Pivot table should look like Table 4.13).

It is important to do this straight away, as by doing so the next step will lead to the valid percent (those who answered the question), whereas when not filtered, it will be the percent (everyone who took part in the survey). Once you have filtered out what is not needed, remember to copy the results from column B and paste it into column C before changing the count into % grand total. The reason to do this is so you can see the valid percent as well as the 'n' so it is easier when writing up the descriptive summary. If you do this, you will be left with the frequency table (Table 4.14).

Table 4.13 Table before percentage grand total

	A	B	C	D
	C14 ▾ ⋮ ✕ ✔ *fx*			
	A	B	C	D
1				
2				
3	Row Labels ⊽	Count of hrprel3	Count of hrprel3	
4	Buddhist	170	170	
5	Christian	20544	20544	
6	Hindu	440	440	
7	Jewish	145	145	
8	Muslim	1232	1232	
9	No religion	12360	12360	
10	Sikh	199	199	
11	Grand Total	35090	35090	
12				
13				

Table 4.14 Table including percentage grand total

	A	B	C	D
	B3 ▾ ⋮ ✕ ✔ *fx*	Count of hrprel3		
	A	B	C	D
1				
2				
3	Row Labels ⊽	Count of hrprel3	Count of hrprel3	
4	Buddhist	0.48%	170	
5	Christian	58.55%	20544	
6	Hindu	1.25%	440	
7	Jewish	0.41%	145	
8	Muslim	3.51%	1232	
9	No religion	35.22%	12360	
10	Sikh	0.57%	199	
11	Grand Total	100.00%	35090	
12				
13				

Two's company in MS excel

Building on the last PivotTable where religion was explored, there is just one additional step to take to carry out a bivariate analysis. All that is needed is to drag and drop the second variable into the PivotTable Field. As before, it is 'gender' and 'religion' that are being explored so all that is needed is to drag and drop 'sex' into the column as shown in Figure 4.30:

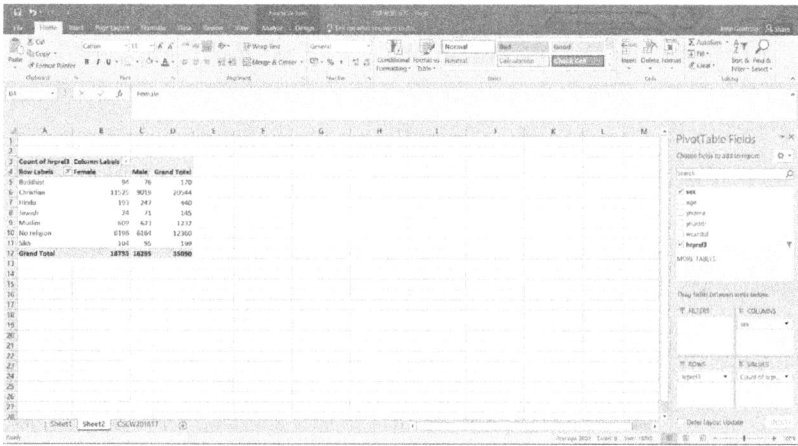

Figure 4.30 Using the Pivot Table function to create Crosstabs in MS Excel

The '*hrprel3*' (Religion) stays in the Rows and Values with the values that are not needed filtered out. As before, the results shown are the count of each value. This can be changed to show the percentage of each value in the same way as before (remember to copy and paste the data to sit alongside the data that will change for ease of comparison):

Step 1: Highlight the results.

Step 2: Right click the mouse.

Step 3: Move down to 'Show Value As'.

Step 4: Select '% of Grand Total'.

The PivotTable should look something like Figure 4.31.

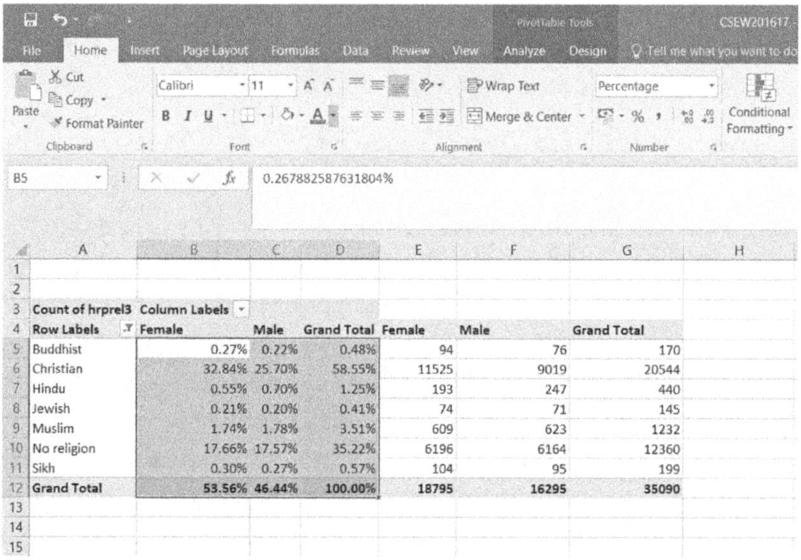

Count of hrprel3	Column Labels						
Row Labels	Female	Male	Grand Total	Female	Male	Grand Total	
Buddhist	0.27%	0.22%	0.48%	94	76	170	
Christian	32.84%	25.70%	58.55%	11525	9019	20544	
Hindu	0.55%	0.70%	1.25%	193	247	440	
Jewish	0.21%	0.20%	0.41%	74	71	145	
Muslim	1.74%	1.78%	3.51%	609	623	1232	
No religion	17.66%	17.57%	35.22%	6196	6164	12360	
Sikh	0.30%	0.27%	0.57%	104	95	199	
Grand Total	53.56%	46.44%	100.00%	18795	16295	35090	

Figure 4.31 Inserting percentages into a Crosstab Pivot Table in MS Excel

Just as when interpreting any table of results, the first task is to simply look over them and begin to build a picture of what the most important trends are. As differences between men and women are being compared, this is the first part of the story already dealt with. The interesting thing to jump out straight away is that there does appear to be differences in religious identity between men and women to varying degrees. It is however, quite difficult to establish. It would be easier to see if the findings were rounded up or down and displayed in a table that allowed them to be compared. As MS Excel does not do this for you (unlike IBM SPSS), it is worthwhile creating a new table that uses all the data. Fortunately, the MS Office suite makes this relatively easy. Table 4.15 is easily made in MS Word whereby all the data from the results is simply copied and pasted into the relevant counts and the rows and columns are given appropriate labels. The rationale for doing this is so that the percent and the frequency can be combined to make the table easier to read. Of course, you might find it easier to do this in MS Excel or by using Google Docs. The process in this instance is less important in favour of the output table that easily displays the appropriate results.

Table 4.15 Making a table in MS Word to show frequencies and valid percent

	Female	Male	Total
hrprel3 (Religion)	% (n)	% (n)	% (n)
Buddhist	0.3% (n = 94)	0.2% (n = 76)	0.5% (n = 170)
Christian	32.8% (n = 11,525)	25.7% (n = 9019)	58.5% (n = 20,544)
Hindu	0.6% (n = 193)	0.7% (n = 247)	1.3% (n = 440)
Jewish	0.2% (n = 74)	0.2% (n = 71)	0.4% (n = 145)
Muslim	1.7% (n = 609)	1.8% (n = 623)	3.5% (n = 1232)
No religion	17.7% (n = 6196)	17.6% (n = 6164)	35.2% (n = 12,360)
Sikh	0.3% (n = 104)	0.3% (n = 95)	0.6% (n = 199)
Total	53.6% (n = 18,795)	46.4% (n = 16,295)	100% (n = 35,090)

The table above displays the results more clearly by showing the percent (%) alongside the frequency (n). Additionally, as long as the blanks and the missing have been filtered out, what will be showing is the valid percent along with the 'n' (frequency or count). Remember, the difference between using IBM SPSS and MS Excel is that you have to make the frequency table when using the latter programme. Of course, this might appear an arduous endeavour, but it is a worthwhile exercise for two main reasons: (1) it is easier for the reader to follow and (2) it also gives you 'flying time' with your data so you become more familiar with the results which will help when it comes to interpreting them. Now that you have a crosstab table using MS Excel, explore the findings and see what jumps out at you. Remember, it is often a case of stating the obvious. It is also worthwhile going through each finding and stating if the percent is higher or lower between men and women. As we are using two variables (bivariate analysis), it is not necessarily a matter of reporting which group had the highest percentage, as while this is interesting and important, it is differences between the groups (in this case differences between men and women) that are of interest.

By exploring the crosstab, we can see that there are a higher percentage (0.3%, $n = 94$) of women who were self-defining as 'Buddhist' than men.

Box 4.7

2-Minute Recap!

Set a 2-minute timer on your mobile phone and 'draw' out the other key findings from Table 4.15 for each religion. Was it men or women who had the higher percentage?

Box 4.8

Time to Get Your Hands Dirty! Interpreting Crosstabs

Using either IBM SPSS or MS Excel (or both if you are being really adventurous) and a data set of your choice, practice running and summarising frequency tables and crosstabs for your data. What interesting patterns do you find?

Looking ahead

In this chapter, we have shown how the seemingly simple statistic of the percentage offers a persuasive way to explore categorical data if used correctly (valid percent showing the frequency). Using percentages, with either one variable (univariate) or two variables (bivariate) allows us to quickly explore, summarise and describe vast amounts of data if using IBM SPSS or MS Excel (or any other software for that matter). It is worth noting that the 'pointing and clicking' that is needed to do this is actually the easy part. What is much harder (but gets easier with practice) is the ability to tell an interesting story with the resultant analysis. In the Chapter 5, we build on this by working with interval-scale and interval-ordinal variables. Again, you will come across some old favourites that you will remember from school. These are mean, median and mode, also known collectively as measures of central tendency (MCT). We will also explore measures of dispersion, something with which you might be less familiar. What is important here is how we use them in the social sciences to provide insights into our data trends.

Chapter Summary

- A key feature of categorical (nominal) data is that it has two or more categories but that they have *no* order or structure to them. Ordinal variables are similar to nominal with a key difference being that the categories can be ordered in some way.
- When we analyse categorical data, we are interested literally in the number (called frequency) of respondents in each category of a specific variable, whether it be 'sex', 'ethnicity', or 'social class'.
- We need to produce frequency tables so that the data can be summarised, examined and analysed.
- EDA requires you to make informed decisions about what can be included but also what can be ignored.
- When presenting findings from frequency tables, as well as reporting the valid percent, it can sometimes be appropriate to merge different categories of responses if it helps develop the narrative.
- When reporting results from frequency tables, it is important to consider how much information is required. Too little and the summary is meaningless; too much and it becomes unwieldy and boring. Be simple and succinct, try and tell a story with what is being reported.
- IBM SPSS and MS Excel can be used to carry out analysis and producing frequency tables; however, these have vastly different processes and functions.

Further Reading

Chapter 4 introduced the use of percentages within descriptive statistics; these suggested readings demonstrate the way that different authors have used percentages as a means to construct interesting statistical narratives.

British Social Attitudes Survey/NatCen. (2019). *Marriage matters.* https://www.bsa.natcen.ac.uk/latest-report/british-social-attitudes-30/personal-relationships/marriage-matters.aspx

This online article (available through the BSA website) primarily uses descriptive statistics, specifically percentages to explore how attitudes towards marriage have changed over time and within different groups in UK society. This article is a great demonstration of how to construct a strong but simple narrative using percentages.

Geary, R. S., Tanton, C., Erens, B., Clifton, S., Prah, P., Wellings, K., Mitchell, K. R., Datta, J., Gravningen, K., Fuller, E., Johnson, A. M., Sonnenberg, P., & Mercer, C. H. (2018). Sexual identity, attraction and behaviour in Britain: The implications of using different dimensions of sexual orientation to estimate the size of sexual minority populations and inform public health interventions. *PLOS ONE, 13*(1), e0189607. https://doi.org/10.1371/journal.pone.0189607

While on the surface, this article is exploring disparities in health, it is actually an exploration into the complex ideas underpinning sexual orientation. Using data from *NATSAL-3*, the article presents a convincing narrative primarily using percentages. Don't let the higher level statistical analysis put you off.

Hick, R., & Lanau, A. (2018). Moving in and out of in-work poverty in the UK: An analysis of transitions, trajectories and trigger events. *Journal of Social Policy, 47*(4), 661–682. https://doi.org/10.1017/S0047279418000028

Another journal article presenting a complex and convincing narrative using little more than descriptive statistics is, in this instance, percentages. This one explores poverty in the UK. While on the surface, the tables look complicated, this is because they are quite large in places. What is important is that as a student, you develop a statistical literacy that will help you cut through all the baggage and find the key narrative.

Joseph Rowntree Foundation. (2016). *UK poverty: Causes, costs and solutions.* https://www.jrf.org.uk/report/uk-poverty-causes-costs-and-solutions

While this report by the Joseph Rowntree Foundation is quite long, it is an excellent example of where descriptive statistics are used persuasively and to make a strong political point, in this instance, the use of percentages.

5

I LIKE BEING AVERAGE, I AM AN INTERVAL VARIABLE

Chapter Overview

Introduction 132

Measures of central tendency and dispersion 132

Averages in everyday life: feedback scores 133

The importance of averages: exploring income and poverty 134

Rich man, poor man, beggar man, thief: the problem when
measuring average incomes 136

What's wrong with being average: income inequalities and the
problem with outliers 137

Measures of dispersion: adding more context to the data 141

Singing stats! 143

Average UK income: but which average? 145

Bivariate analysis using MCT 147

When averages are just plain silly 148

Writing up the results: the end is nigh 149

'How to' calculate MCT and MoD using IBM SPSS 151

'How to' calculate MCT and MoD using MS Excel 155

Looking ahead 155

Further Reading 157

Introduction

In this book, so far, we have identified that there are different types of data based on what aspect of social life we are trying to measure. We should also now appreciate that how we clean, analyse and summarise that data differs. If you are reading this chapter linearly, then you know that in Chapter 3, we looked at different types of variables (and data) and in Chapter 4 we looked specifically at categorical variables (and data). If you are dipping into this book, then we presume that you know all this and can jump straight into this chapter which looks at interval variables (and data). If you don't, then you may want to stop and do some reading first.

If you look back at Chapter 3, you will recall that the *average* weekly salary for different age groups was presented, with 16- to 17-year-olds earning the least amount of money per week. Did any amongst you wonder which average was being referred to (the mean, median or mode)? We bet you didn't because most people don't really consider this when they see the word 'average' bandied about in the media, but we hope that by the end of this book you become more critical and reflective. Averages are funny old things in that we don't have just one, we have three. Have you ever stopped to think why this might be? One of the aims of this chapter is to explain and explore the reason behind this and to hopefully show you just how useful averages are to us as social researchers. They are important and each do a very different job in explaining data, so it is important to understand the difference between each type. Averages are a very popular form of descriptive statistics. Their strength comes from allowing us to aggregate, and make sense of, large amounts of scale data. In this chapter, we are going to continue our exploration of univariate analysis (using one variable) looking at **interval variable** or, as it is sometimes called, continuous data. As highlighted in Chapter 3, interval data can either be ordinal (ordered such as a Likert Scale, 'strongly agree' to 'strongly disagree') or scale (continuous data, such as, age, weight, height or hours spent on social media).

Measures of central tendency and dispersion

There aren't many good stats jokes but here's one for you: *'Measures of central tendency have a bad reputation of being very uncaring. But you know what, they're not all mean!'* When we refer to **measures of central tendency** (MCT), we are in fact referring to any of the three averages: the mean, the median or the mode. These are the same averages that you learnt in primary school, so even though the phrase 'measures of central tendency' sounds horrible (and it is, we think it sounds like some sort of medieval punishment) just remember that at the heart of MCT are the averages that you learnt in school, which we hope are less horrible. Of course, you may not have heard of measures of dispersion (MoD, which also sounds like a punishment of some kind), but you probably covered this in high school maths. Because we think you have possibly blotted out high school maths from your memory, MoD means simply how spread out the numbers are within a data set.

To the newly enrolled social science student on an undergraduate statistics course, finding out that they have to work with averages *again* has resulted in many an eye-roll. For many, they thought they had left maths behind at school. We have had many students simply ask, 'Why are we doing this' (often with a sense of fear showing in their eyes). We have heard cries of *'I've not come to university to do maths!'* (angry voice and face ☹). Yet statistics such as MCT tell an important story when using descriptive statistics. It can be quite some task trying to persuade a resistant social science student of their utility. It is also important to note that none of this is technically mathematics but arithmetic but we're guessing, you don't care! However, averages are used in our everyday life and perhaps we do not always notice them? We will go into more detail about the different averages below, although there was a short recap in Chapter 1 of this book that may have helped jog your memory about what they are and how they are different.

Averages in everyday life: feedback scores

Let's think about how we use averages in our everyday lives to help us make decisions. Think for example, about when you download and rate an app that you buy from an app store, or when you are planning on buying something from a website like eBay. These also work on averages. If the maximum feedback rating allowed is 5 stars, then the highest star rating anyone can receive is 5. Similarly, if the lowest feedback score that can be given is 1, then clearly, 1 will be the lowest score anyone can receive. We therefore know that the popularity of the app will fall somewhere between 1 and 5 stars. It might be that an app has a rating of 5 stars, strongly suggesting it is popular with those who have already used it. In essence, it is 'above average'. Yet another important piece of information that can help you make this decision is also needed. These types of feedback scores are heavily influenced by the number of ratings that are received. One feedback rating might have top marks of 5 yet might have only had one person provide a review. This is not to suggest that the rating is dishonest. Rather, there have not been enough people providing feedback to get an average score to make an informed decision. If however, an app has a lower average feedback rating of 4.3 stars but has had more than 1 million reviews, it would be safe to assume that at least for these users, they collectively 'like' the app. You know this intuitively because you look at these sorts of ratings and make decisions about products based on them all the time. What would happen to the rating if 500 buyers rated the app as 1, whereas a different 500 reviewers gave it the full 5 marks? The mean average would be 3 (all the scores added together and divided by the count). Can you think what the median average would be? Median is the middle number. We'll leave that there for a short time but return to it later in the chapter. What is important to keep in mind is how they are used in the social sciences, and how this might differ from what was taught at school. The key difference here is how they are used to explore and interpret interval/scale variables in a data set. Let's look at two examples using income again.

Box 5.1

2-Minute Recap!

Set a 2-minute timer on your phone and answer both questions below. Remember to write the answers down.

What are the three different types of averages?

What are the differences between them?

The importance of averages: exploring income and poverty

The quote below is from the Joseph Rowntree Foundation exploring income levels in families with children:

Poverty rates are consistently highest among children and their parents. Twenty years ago, a third of children lived in poverty. This fell by 15% between 1994/95 and 2004/05, to 28% of children. The child poverty rate fell to its lowest level (27%) in 2011/12, but has started to rise again in recent years, reaching 30% in 2015/16. (UK Poverty, 2017, JWF)

You might be wondering why the descriptive statistics being used are percentages rather than averages which is the focus of the current chapter. Think for a moment though: The reason we know the above percentage and changes in poverty levels year on year is because the average income level has been collected and calculated. The key to the data is an average: in this case, the average income and the percentage of families and children that fall below this average to the extent that they are considered to exist in a state of poverty. We will come back to average salaries later. Let's look at another example, the gender pay gap. We know that for every £1 a man earns, a woman will, on average, earn around 80p for doing an equivalent job (Full Fact, 2015). We know this because some curious researchers have kindly collected and analysed such data so that we can further explore gender inequalities throughout UK society. It would be too difficult to observe all the different incomes of men or women throughout the UK as this would clearly be too many for us to make sense of. Instead, we can use the averages to summarise the data and give a single or range of figures from which we can work to develop a story of gender inequalities. Here, the average is the difference between what men and women earn. Women will, on average, earn 20p less than their male counterparts. Similarly, the UK's ONS (2018) reported that in 2017, the average hourly pay for a man was £14.48, while for a woman it was £13.16. Whichever evidence is used, both clearly show that on average, women in the UK earn less than men. Therefore, descriptive statistics, whether it be averages or some other MCT, become crucial in our preliminary understanding of levels of families living in poverty or the gender pay gap. Without these statistics, we would not be able to identify or define the problem itself – how can you know there is a pay gap unless you know the average pay.

It almost seems disingenuous to leave out how we are sure about the numbers living in poverty or the gap in earnings between men and women if we don't acknowledge the data that supports such claims. It might make the novice researcher wonder what other important theories are premised on such quantitative evidence. Crucially, MCT used by the social sciences are not abstract numbers but, rather, provide important (some may say vital) evidence about a particular sample and/or population. Let's continue the theme of income and explore the strengths and weaknesses of MCT when measuring wealth differences within the UK population.

Box 5.2

Pause for Thought

Think about the number of times where averages are used in our everyday life. Once you have a few, think about how and what they are trying to explain in ways that 'raw' data would simply not do. You could do this with some classmates and see how many you all identify.

Rich man, poor man, beggar man, thief: the problem when measuring average incomes

Did you know that as of 2018, the ONS reported that the UK average income is £27,300 per annum? This is a great headline figure, although it only tells a partial story. Knowing the average income is an important economic indicator that not only shows the wealth of individuals and families but also helps us determine what would be considered as 'living below the poverty line'. However, social science researchers and students need to know so much more to make sense of this figure. For example, which MCT is being used? Is this gross, net or disposable income? We could also ask if this is the average income of an individual or a family/household. Hopefully, all these questions will be answered below. But first let's jog our memories and review what we learnt back in primary school about averages.

Mean (is average): is what most people would consider *the* average and as discussed, it is the arithmetic average. To calculate the mean, add all the numbers together (sum them) and then divide by the total number of observations. For example,

7 + 9 + 5 + 2 + 8 + 6 + 7 + 12 + 7 = 63 divide by 9 = 7 – *The mean is 7.*

Median (is middle): to calculate the median requires the data to be ordered from low to high and the number that falls in the middle is considered the median average. If however, there is no middle number (this happens when you have an even number of scores), then the two middle numbers are added together and divided by 2 to give the median average. For example,

7, 9, 5, 2, 8, 6, 7, 12, 7 would need reordering from low to high:

2, 5, 6, 7, 7, 7, 8, 9, 12. The middle number when ranked low to high is 7 – *The median is 7.*

Mode (is most): to calculate the mode we must establish which number occurs the most often. We need to remember that there can be more than one mode in a distribution of scores or data; two modes is known as 'bimodal' and more than two modes is known as 'multimodal'. Using the same numbers as used above:

2, 5, 6, 7, 7, 7, 8, 9, 12. The number that occurs most often is 7 – *The Mode is 7.*

As you can see with the three examples above, although the same scores are used in each example, the averages are each calculated differently. For our scores, all our averages are the same which is a good sign (more of that later) but your averages may all be different depending on the data that you have.

Box 5.3

2-Minute Recap!

Set a timer on your phone for 2 minutes and calculate the mean, median and mode for the following data (remember to rank them for median and mode):

4, 6, 7, 5, 6, 2, 12, 6

Mean =

Median =

Mode =

You might wonder if having three different types of averages are necessary. As will hopefully become clear later in this chapter, the point of having three different ways of calculating averages stems from the shortcomings of each type and how they can produce unrepresentative, meaningless or even misleading results. Let's explore this further using average income.

Box 5.4

Reflective Exercise

Many societies, including the UK, have huge socio-economic inequalities where some people have a lot of money while others have very little or none. With this in mind, which average mean, median or mode would be most suitable to calculate the average income?

What's wrong with being average: income inequalities and the problem with outliers

One of the main issues to take into consideration when measuring average incomes is the huge differences in incomes that exists in the UK (and across much of the world too!). If we used the mean average to calculate average income, the huge extremes of different income levels would make the results meaningless and not representative of the UK population. This is because we have people in the UK who have little or no income compared with some who have billions of pounds in the bank. Of course, most people have more than nothing but less than a billion. The problem here is that the mean average is susceptible to extremes which are known as **outliers**. Outliers effectively skew the data and therefore the results. In terms of extreme differences, just consider the difference between having no money/income and having several billions. Firstly, look at how many zeros are needed in a billion:

0

10

100

1000

10,000

100,000

1,000,000

1,000,000,000

Look at this another way. One million seconds relates to 1.65 weeks; however one billion seconds is the equivalent to 31 years and 7 months. Look at that in terms of income and it becomes apparent that someone with so much income is going to severely mess up the average UK income such is the extent of this as an outlier. Let's explore outliers with some simple 'made up' data ('hours spent online shopping') to illustrate this point. Table 5.1 has four columns of very different data. The mean average is calculated at the bottom.

Table 5.1 Hours per week online shopping

Cell	A	B	C	D
1	2	0	5	5
2	3	1	6	6
3	4	1	7	7
4	5	1	5	95
5	6	1	6	6
6	7	1	7	7
7	8	1	5	5
8	9	1	6	6
9	10	47	7	7
Mean average	6	6	6	16

As you might remember, the method to calculate the mean is to add up all the numbers and divide them by how many numbers there are. Looking at column A in the table above, to calculate the mean average then:

$$2 + 3 + 4 + 5 + 6 + 7 + 8 + 9 + 10 = 54/9 = 6$$

But is 6 hours online shopping truly representative of the data? It does not appear so. There are no obvious outliers but the data is quite spread out.

Now look at column B:

$$0 + 1 + 1 + 1 + 1 + 1 + 1 + 1 + 46 = 54/9 = 6$$

As can be seen, the mean average for this column's data is also 6 hours. Again, we have to ask if the result (6 hours) seems representative of the actual data and again we would have to say no! Yet it has produced the same result as column A but has very different looking data. This is the result of cell B9 (47) that could be considered an outlier. If the point is to summarise the data, then the results from columns A and B do not look like they are doing a good job.

Let's look at column C:

$$5 + 6 + 7 + 5 + 6 + 7 + 5 + 6 + 7 = 54/9 = 6$$

Again, there is an average of 6 hours, and in this instance, the result do more accurately summarise the data. The data is not too spread out and nor does it have any outliers.

Lastly, let's look at column D:

$$5 + 6 + 7 + 95 + 6 + 7 + 5 + 6 + 7 = 166/6 = 16$$

Columns C and D are very similar with just one difference; 95 in column D that is a pesky outlier. However, look at the impact this has on the result. Again, it seems to bear little resemblance to the actual data. This is the problem with using the mean average in that it not only 'smooths' the data it is summarising, it is also heavily influenced by extreme scores. Moving back to the example of average UK income, column B is likely to be the most representative of income levels. The fewest people having either very little at one extreme, or an awful lot of income at the other. Most however, fall somewhere in between. This leaves us with a problem of how to best calculate average incomes. The answer, of course, is that we turn to a different MCT that would better summarise the data and would be less susceptible to such extremes: the median or the mode.

Now using Table 5.2 (which repeats the data from Table 5.1), calculate the median and mode (for the answers, see Table 5.3, but remember, don't cheat, give it a go before looking!).

Table 5.2 Hours per week online shopping (with MCT)

Cell	A	B	C	D
1	2	0	5	5
2	3	1	6	6
3	4	1	7	7
4	5	1	5	95
5	6	1	6	6
6	7	1	7	7
7	8	1	5	5
8	9	1	6	6
9	10	47	7	7
Mean average	6	6	6	16
Median average				
Mode average				

Note. MCT = measures of central tendency.

Table 5.3 Hours per week online shopping (with results from MCT)

Cell	A	B	C	D
1	2	0	5	5
2	3	1	6	6
3	4	1	7	7
4	5	1	5	95
5	6	1	6	6
6	7	1	7	7
7	8	1	5	5
8	9	1	6	6
9	10	47	7	7
Mean average	6	6	6	16
Median average	6	1	6	6
Mode average	(No mode)	1	5, 6, 7	6
Min number				
Max number				
Range				

Let's explore the results (once you have calculated them). Table 5.3 shows the results for Median and Mode average.

In column A the mean and median are the same result (6 hours). The median average therefore does not provide a better summary than the mean. There is no mode as there are no repeating numbers.

Column B has a mean of 6 hours, a median and mode of 1 hour. Looking at the data, they do seem to better summarise the figures than the mean average. In this example then, perhaps it would be better to use either the median or the mode to summarise the number of hours spent shopping online.

Column C has a mean of 6 hours, a median of 6 hours and a mode of 5, 6 and 7 hours (it is therefore multimodal). Again, both the mean and median average do seem to summarise the data quite well. The mode however, while representative of the data, offers us nothing in the way of a summary.

Column D has a mean of 16 hours and a median and mode of 6 hours each. Again, median and mode summarise the data much better than the mean average.

Therefore, when summarising trends in data which tend towards extremes, either low or high, such as income, house prices and suchlike, it is usually better to use the median or the mode; although typically, it is the median that is used because of the tendency for there to be more than one mode in a large spread of numbers.

Measures of dispersion: adding more context to the data

When deciding which MCT to use, it is helpful to be able to see the data, as we could with our example of hours per week online shopping. We were able to see that column A was very different from column B, and that the results in columns B and D were influenced by outliers. It's easy when there is only a small amount of data; likewise, we can easily manually calculate our averages without needing statistical software. However, what would happen if we needed to make the decision of which MCT to use when there are simply too much data to observe? Even if a data set only had 100 responses, this would be a lot to work through to get a feel for which MCT was best to use. Yet many data sets have several thousands of responses making it impossible to observe and get a feel for the data; think back to Chapter 2 and how large the GSS and Natsal-3 data sets were. So, what could we do to help us make a decision? One way would be to look at the spread of our data, using Measures of Dispersion (MoD) take the data and rather than summarising which figure represents the middle, they instead produce a summary of the diversity of the data. There are a number of different statistical techniques that are used to reveal how dispersed the data is with the two main ones being the range and standard deviation.

The simplest MoD to calculate is the **range**. The range is an MoD whereby the smallest number (the minimum, typically abbreviated to min.) is subtracted from the highest number (the maximum, typically abbreviated to max.). Have a go at calculating the range for Table 5.3.

You might of course be asking why we need to keep making this more and more complicated. But remember, the reason we do this is to help us develop a narrative using quantitative data while also accounting for some of the inherent deficiencies of using MCT. When we are exploring MCT, it is useful to know how spread out the data is as this should indicate how representative the mean is of the data. This depends on what was being measured. If, for example, the age of the UK population was reported, we should expect a very high range as long as the youngest and the oldest persons are included. Currently, the oldest person in the UK is Grace Catherine Jones who was born on 16 September 1906. As of 2018, this would give us a range of 112 years as the youngest person (a newborn baby) would be zero years old. In this example, Grace could be considered an outlier (sorry Grace) as the average life expectancy for women in the UK is around 82.8 years as measured by the ONS, between the years 2013 and 2015. The mode figure is, however, slightly higher with the most common age at death for women being 89 years. In this instance, we would not use the median average as this would not be representative of the average age people in the UK would live to. So, for some variables, we could expect to find a high range (like age or income), but for others (the heights of all the children in a class-room of 5-year-olds), we might find a small range. Nevertheless, a high range is a use-ful means to determine whether there might be outliers that could be influencing our mean, making it less representative. If we look at Table 5.4, it now shows the spread (dispersion) of the data in each column. The bigger the spread, the higher the range.

Table 5.4 Hours per week online shopping (with results from MCT and MoD)

Cells	A	B	C	D
1	2	0	5	5
2	3	1	6	6
3	4	1	7	7
4	5	1	5	95
5	6	1	6	6
6	7	1	7	7
7	8	1	5	5
8	9	1	6	6
9	10	47	7	7
Mean average	6	6	6	16
Median average	6	1	6	6
Mode average	(no mode)	1	5, 6, 7	6
Min number	2	0	5	5
Max number	10	47	7	95
Range	8	47	2	90
Std Deviation	2.73861	15.37856	.86603	29.63528

Note. MCT = measures of central tendency; MoD = measures of dispersion.

Look at column A. It has a minimum number of 2 hours and a maximum num-ber of 10 hours, which gives us a range of 8 hours. If we knew that when we first calculated the mean, we might be sceptical of the mean average of 6 hours spent online shopping.

Similarly, if we knew we had a range of 46 hours in column B and a mean average of 6 hours, it might again suggest a degree of caution with the suggestion that the mean average should not be used here. It is column C that has the smallest range suggesting that the data has the least amount of spread or smallest differences. It does look like the mean average is more representative of the data. Column D has a large range of 90 hours, which again suggests that we should treat the mean average with some scepticism. So including the range in the exploration of the data can help better explain it and see if there are any issues that need to be acknowledged. However, using the range does have its limitations. For instance, take the example of age above, having a large range is actually okay in this instance and doesn't really reveal just how the data is spread out. What would help is a standardised approach that would show more clearly just how dispersed the data is. This is called standard deviation.

Singing stats!

Before we look at standard deviation, we wanted to offer a simple way to remember the differences between the averages and the tange: to sing them! Using the tune of Frère Jacques (all together now ♩):

Mean is average, mean is average

Mode is most, mode is most

Median is middle, median is middle

Range high low, range high low.

Perhaps there's a potential genre of music to be made 'singing stats' . . . who knows!

Standard deviation: horrible name, but really useful

As the name implies, **standard deviation** is a standardised approach by which the spread of the data is reported. Using the range is less precise when comparing data from different sources. Think for example, if we had data with a minimum of 100 and a maximum of 190. This would give a range of 90. Now compare that with data that has a minimum of 1000 and a maximum of 1090. That too would have a range of 90 yet the data would clearly be very different. This is why we need a standardised approach by which to compare interval data; the most commonly used is the standard deviation. Calculating standard deviation, however, is considerably more complex than calculating the range and requires each score to be subtracted and squared from the mean using a specific equation. You can go online and find out easily how to do this manually, which works okay if you have a small data set. Alternatively, you can use a statistical software to do the hard work for you, which let's face it makes life easier, especially when we have very large data sets. There are IBM SPSS and MS Excel 'how to' guides at the end of this chapter. At this point, how to calculate standard deviation is less important than understanding it and knowing how to interpret and use it.

In providing a standardised approach that explores the spread of data, the focus is on the spread itself and not just what is being measured. In essence, what we are doing with standard deviation is asking 'How much does the data vary from the mean average?' This allows us to assess whether we should trust the mean as truly representative of our data's average or midpoint. As such, a low figure for standard deviation would mean that the numbers are quite close to the mean (not spread out too much). That would mean that a large standard deviation indicates a large dispersion and so differ quite a lot from the mean average. Look at Table 5.4, which again shows the number of hours per week online shopping. The standard deviation is at the bottom.

Looking at column A, there is a mean of 6 and a standard deviation of 2.73861. If compared with columns B and D, the standard deviation shows that the data is not too spread out. Column B has a larger standard deviation, of over 15, and look, the data is very spread out. However, looking at column D, there is a standard deviation of over 29 and look at the data; it is the most spread out. Compare them with column C which has a standard deviation of under 1 (.86603) and you can see that this has the least spread out data.

Home on the range? using the range rule for a 'quick' standard deviation

Calculating the standard deviation can be difficult, especially if you are doing it manually. However, there is a quick method that you can use to calculate an approximate standard deviation, which uses the range. This is called the **range rule** whereby the range is simply divided by 4. Take another look at Table 5.4 and carry out the range rule calculation and then compare it with the actual standard deviation. They are in the same ball park so provide some idea of what the standard deviation will be. However, a word of caution here. The range rule produces an approximate standard deviation and is only a suitable approach when first eyeballing the data and doing very early exploration of it, as discussed in Chapter 2, but it should not be relied upon in final analysis or summaries. Perhaps the times to use this method of calculation is when there is no access to a statistical software and a deeper exploration of the data would be beneficial. However, in most instances, it would be better to let the statistical package of choice do the work so that the actual standard deviation is calculated.

Box 5.5

2-Minute Recap!

Put the timer on your mobile phone for 2 minutes and then do the following.

In the first minute, write down the difference between the range and the standard deviation.

In the second minute, calculate the standard deviation using the range rule for the data listed below:

36

92

27

59

Average UK income: but which average?

What we have covered thus far has explored what the different types of averages and MoD are. Now reconsider the example of income to see how they can all tie together. The question that has not yet been answered is which MCT should be used when explaining average UK income. The pros and cons have to be taken into account with the acknowledgement that all might have some shortcomings. The mean average has already been discounted as it is too susceptible to extremes (outliers). This leaves the median and mode. Before reading on, try to consider which you think would be the most appropriate MCT to calculate average income in the UK. Then read on and we will explore some more.

Let's first consider the mode: unlike the mean, the mode average is not suscep-tible to extremely high or low outliers. It is also quite easy to calculate. Perhaps this would be the best measure to use to calculate average income. The advantage of the mode is how it is the most recurring number or figure. If, for example, most people earned £20,000 a year, this result would be representative of most of the population. However, if we needed a single figure from which to calculate, for example, what it was to be living in poverty, then the mode would not always produce a single figure (we could have data which is bimodal or multimodal). We could, therefore, have instances where the mode income is £20,000, £25,000 and £30,000 meaning the result is multimodal, which doesn't sound good and isn't.

What about the median then? Like the mode, it too is less susceptible to extreme outliers. The median average uses the number which is found in the middle of the data, which makes it a good candidate to be our measure for average income. One problem that might arise, however, is how the result might not actually be a figure in the data if, for example, there is no middle figure and the two middle data need to be used to calculate the median. Think back to the example above where we had 500 people rating an app at 1, whereas a different 500 people rated it at 5. The mean average was 3 but so too was the median. That is because there is no middle number, which means the two middle numbers (1 and 5) are added together and divided by 2 (1 + 5 = 6/2 = 3). Think for one moment however. What if 499 people rated the app at 1 with 501 giving it the full 5 stars? That would again mean there is no middle number. The two middle numbers are 5 and 5. You do the maths!

The issue here is that all of the averages have advantages and pitfalls. A decision has to be made as to which MCT better summarises the data, in this instance, average income. Do you have a hunch? To summarise:

Mean – susceptible to extreme outliers

Median – might not actually produce a result that is found in the data

Mode – might produce more than one result

Did you guess it was median? This measure is used by most offices of national and official statistics to calculate the average income of populations. By using the median, the issue of outliers is resolved as is the issue of potentially having more than one result. Moreover, by using the median average, it means that there are as many that fall below the average as who fall above it. It is with such a result that makes it possible to calculate what is considered to be living 'below the poverty line'. Of course, this is a subjective matter that often depends on governmental policy determining what income is deemed to fall below the poverty line. The point is that the data is sufficiently summarised that it usefully represents the population and provides an accurate estimate of what the average income is for a UK citizen. The current UK median average household disposable income is £27,300 per annum.

Box 5.6

Reflective Exercise

As wages go up and down, year on year, the figure presented here is taken from 2018. Why not go over to the website for the ONS and see if there are any more up-to-date figures. You could also search for the average income for other countries and compare with the UK.

Bivariate analysis using MCT

We have been focusing on one variable, average income, which means we have been conducting univariate analysis, specifically using the median average income for the UK population. However, as we showed in Chapter 4, it can also be useful to carry out bivariate analysis, by simply adding an additional variable. We already have the income level for the UK and we know that men and women have on average different hourly pay, with women typically being paid less than men shown in Table 5.5.

Table 5.5   Median average hourly pay for men and women between the years 2011 and 2017

Year	Men's median hourly pay	Women's median hourly pay
2011	£13.12	£11.75
2012	£13.27	£12.01
2013	£13.60	£12.24
2014	£13.61	£12.30
2015	£13.85	£12.51
2016	£14.16	£12.82
2017	£14.48	£13.16

Note. Adapted from Annual Survey for Hours and Earnings.

Table 5.5 shows the median average hourly pay for men and women from the years 2011 to 2017. It shows that there is an upward trend in pay for both men and women, but that men are consistently paid more than women in every year. Figure 5.1 shows the same data as above.

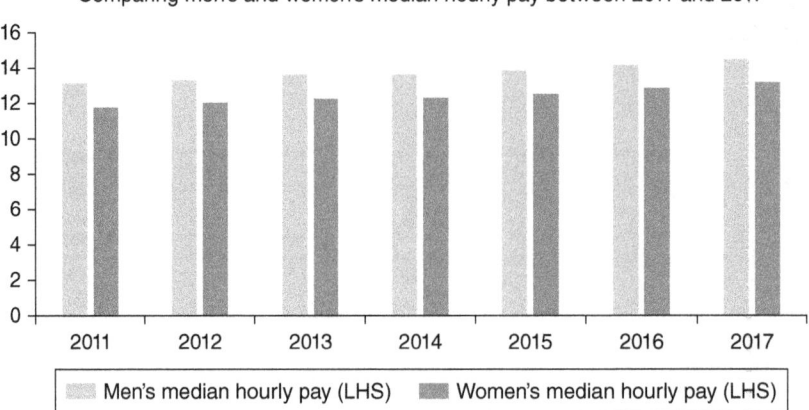

Figure 5.1 Median average hourly pay for men and women between the years 2011 and 2017

Source. Adapted from 'Understanding the gender pay gap in the UK (ONS 2018).

Again, it is the median average that is used to show the average hourly wage. They show that consistently men are paid more than women, but that the pay gap is getting smaller. When we conduct bivariate analyses, the output boxes we get can be quite large and it can be easier instead to visualise the data using boxplots, Q:Q plots, or bar charts; Chapter 6 will show you how to do this.

Box 5.7

2-Minute Recap!

Set the timer on your phone for 2 minutes and see how much you can remember what the strengths and weaknesses are of mean, median and mode averages.

When averages are just plain silly

Using the example of income demonstrates why we need different MCT, but we only need to look at ourselves and our families to know just how difficult it is to appropriately summarise some forms of data. For example, news reports often report the average number of children per family; using the mean average at first glance is not very useful or representative. For example, on the BBC webpage of 15 October 2018, the headline states, 'Women have 1.9 children on average, a record low' https://www.bbc.co.uk/news/health-42110846.

Are they (and all other media organisations) correct to use this type of average? Well think about it for a moment and consider what would happen if either the median or mode was used. Remember, median is the middle figure and mode is the most occurring. To explore this further, think also about the current context of what constitutes a family. As reported on the ONS website, 'In 2017, families with no children or without dependent children were more common than families with dependent children' (ONS, 2017). As not all families have children, we could be left with a median average of zero (the one in the middle). If we used the mode, or the most occurring number, and remember that not all families have children, zero might be the most recurring number. When we look at it like this, we might conclude that 1.9 children per family is in fact more informative and representative than the zero we would get from median and mode.

'I am above average actually!'

There are also times when most of us are 'above average'. Yet how can this be possible; how can the majority of people be above average – this simply should not happen! The reason why this is, however, is when we start counting things like our legs, eyes or ears. Most of us have two of each. Some of us might have one, and some have none. No one has three legs, eyes or ears. So this means that the average number or legs, eyes or ears must be below two, making most of us above average! Here it would be appropriate to use either median or mode as that would then give us a more representative summary of the number of limbs we have which is two. It is for this reason that when MCT are used, there is a need to include additional information that is explored below.

Box 5.8

Pause for Thought

Just how average are you? Think for a moment about all the differences and similarities you have with other people in your class. It might be the size of your family, the amount of time you spend on the internet or the distance you need to travel to school, college or university. Can you make a list of these and share it with someone else.

Writing up the results: the end is nigh

Now that you know the strengths and shortcomings of MCT and MoD, you would be forgiven for thinking this is the end of the story. Alas, sadly not, although we are quite close to the end point of how to use them. The last job that needs to be done is to learn how to report the findings you have calculated. To learn this, it is important to also learn how to avoid common mistakes when reporting the results from these statistics. At one level, it is again a simple matter of *stating the obvious* such as '*the mean average was 1.8 with the median of 2 and a mode of 2. The standard deviation was .2 and the range 1*'. You might ask 'Where's the story in that?' and you would be correct. One way to add to the narrative here is to put in what was being measured. Was it the average feedback rating of an eBay seller or the average number of legs of people living in the UK (think back to the example earlier). It would be much more informative if we put '*the average number of legs in the UK population is 1.8 with the median of 2 and a mode of 2. The standard deviation was .2 and the range 2*'. This can be accounted for by the fact that most people have two legs, some people have one or no legs whereas no person has three legs. Therefore, in this instance, all that is required is that you are explicit in what is being measured within the write-up.

There are other mistakes that seem to be repeated, especially in the media, that should be avoided. For instance, if you look at the BBC webpage reporting gender pay inequalities, once at the webpage, we are invited to type in an organisation whereupon we are given the findings stating if there is, or is not, a pay gap between men and women. The median average is reported along with the following statistics:

- 78% pay men more than women
- 14% pay women more than men
- 8% have no gender pay gap

(https://www.bbc.co.uk/news/business-43632763).

We typed in Manchester City Football Club (MCFC) and Manchester United Football Club (MUFC; the two teams we follow). Here is the headline figures.

Manchester City Football Club:

The average man at this company is paid 17.2% less than the average woman.

It is interesting to see how this company is one of the 14% that has the reverse gender pay gap in that it is men who are paid less. Take a close look at the data again and see if you can identify an error. To add a little balance, below is the same data for MUFC:

Manchester United Football Club:

The average man at this company is paid 7.4% less than the average woman.

Did you identify the issue with the way this average is being reported? Well first of all, you did read it correctly as it does state that both football clubs pay the average man *less* than the average woman. It is not the type of average used as the median is correct. Rather, it is what is being reported as the average that is incorrect. The narrative suggests that 'The average man at this company is paid 17.2% (MCFC) or 7.4% (MUFC) less than the average woman'. But what is meant by an 'average man or average woman'. Is this someone who is 18 years old and works as a steward on match days or the chief executive? The issue here is not the data but the way they have reported the wrong average. A better way to express this would be to say that '*the average salary of men in this company is 17.2% (MCFC) or 7.4% (MUFC) less than average salary of women*'. There are many such mistakes littered throughout the media where the headline seems to take precedence over the ability to report the correct average (in the example above, the correct average is salary, not being a man or a woman). This will be dealt with in much more detail in Chapter 7 of this book.

Box 5.9

Reflective Exercise

The mainstream media have a tendency to misreport statistics, such as averages, often suggesting the person is the average rather than the actions of the person. They might suggest that the average woman spends such and such on cosmetics, or the average man is more likely to speed when driving. In both these instances, it is the spend on cosmetics and the speed of driving that is the average, not the person. Go online and search for how MCT are reported, and if you find some that are not correct, rewrite them as they should be.

'How to' calculate MCT and MoD using IBM SPSS

Using the same data set as used in Chapter 4 (CSEW 2016–2017), it is a similar process when using IBM SPSS. There are, however, a few more actions that you need to click to produce the appropriate output. Follow the below steps and accompanying screenshots. The first four steps are the same as when using categorical data, which we did in Chapter 4. Figure 5.2 shows the first three steps:

Step 1: Analyze

Step 2: Descriptive Statistics

Step 3: Frequencies

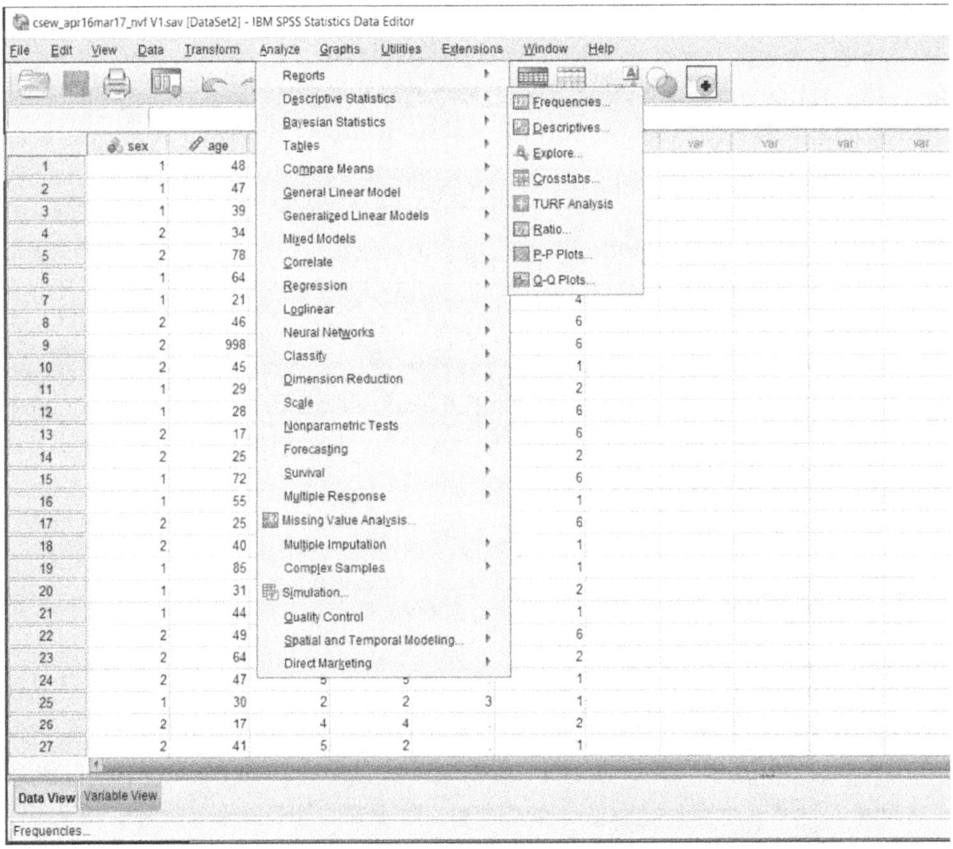

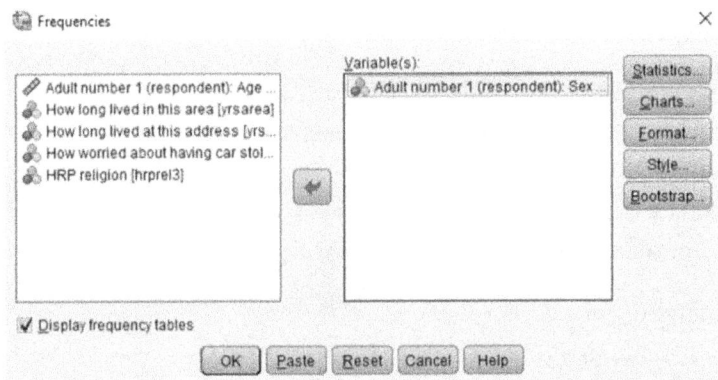

Figure 5.2 Accessing Frequencies using Analyze on the IBM SPSS toolbar

Step 4: In the new dialog box, Figure 5.3, drag *Adult Number 1 (respondent): Age [age]* from the left-hand window to the right under *Variable(s)*

Figure 5.3 Selecting a variable in the Frequencies dialog box, IBM SPSS

Step 5: Click on 'Statistics'. A second dialog box will appear (Figure 5.4)

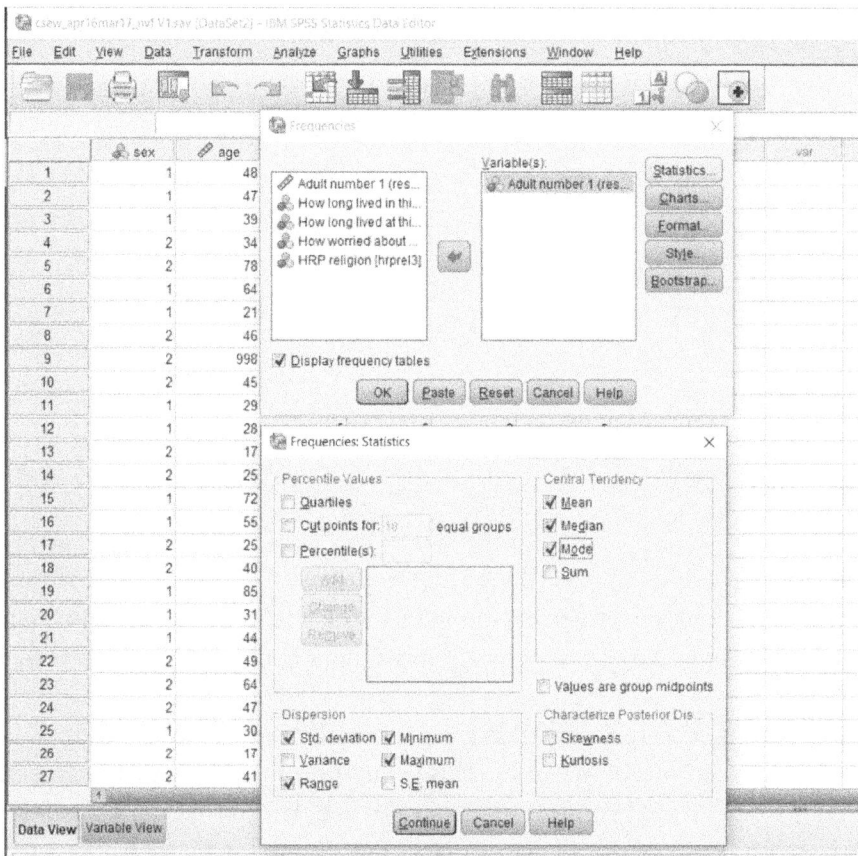

Figure 5.4 The Statistics dialog box in Frequencies, IBM SPSS

Step 6: Tick all of the following:

Mean

Median

Mode

Minimum

Maximum

Std Deviation

Range

Step 7: Click 'Continue'

Step 8: Untick Display frequency table (from the Frequencies dialog box)

Step 9: Click OK

The following output table (Table 5.6) will be produced.

Table 5.6 IBS SPSS output showing the MCT and MoD for respondent's age

Statistics		
Adult number 1 (respondent): Age		
N	Valid	35284
	Missing	136
Mean		51.30
Median		51.00
Mode		45
Std. Deviation		18.505
Range		84
Minimum		16
Maximum		100

Note. MCT = measures of central tendency; MoD = measures of dispersion.

As with the previous chapter, there is a lot of detail here, all of which is important. Therefore, don't rush when exploring this, look at both the words and the numbers. Before reading on, think about all of this and question what the table is actually telling you.

The first thing of note is what this table is actually measuring. It is easy to overlook this and focus on the statistics. This would be a mistake, and as stated previously, it is essential that you add the appropriate detail so that the data all begins to make sense. The key word here is 'years'. Reading the table then shows us that the mean age is 51.30 years with a median age of 51 years. The most common age is 45 years, so slightly lower than the mean and the median. There is a range of 84 years with a standard deviation of 18.505 years. The minimum age is 16 years and the maximum age is 100 years. There are 136 missing pieces of data where age was not recorded. This meant that the data came from 35,284 respondents. As this data is measuring age, it would be expected to have a large standard deviation. Of course, we might want to see if any of the results would benefit from being rounded up or down. For instance, the mean age of 51.30 years would not lose any of its explanatory power if it was rounded down to 51 years. Similarly, the standard deviation of 18.505 years could be rounded down to 18.5 years, and again, nothing is really lost. However, there are no hard and fast rules about rounding up and down. We therefore suggest you check with your tutor before making this type of decision. And when in doubt, don't! Simply report the actual results (better safe than sorry).

'How to' calculate MCT and MoD using MS excel

As indicated in Chapter 4, using MS Excel to calculate descriptive statistics is quite different and requires a small element of writing formula. This might be quite new to you, and if so, it might be beneficial to find and watch some online videos that will help with this. As we want to find the same statistics as we had with IBM SPSS, we need to know the formula as shown in Table 5.7 for the following.

Table 5.7 MS Excel formula needed to calculate MCT and MoD using CSEW 2016–2017

MCT/MoD	MS Excel Formula	MS Excel Formula Using CSEW 2017–18	Results
Mean	=AVERAGE	=AVERAGE(B2:B35421)	51.30
Median	=MEDIAN	=MEDIAN(B2:B35421)	51
Mode	=MODE	=MODE(B2:B35421)	45
Maximum	=MAX	=MAX(B1:B35420)	100
Minimum	=MIN	=MIN(B2:B35421)	16
Standard deviation	=STDEV.S	=STDEV.S(B1:B35420)	18.505
Range	[a]	=B35425-B35426	84

Note. MCT = measures of central tendency; MoD = measures of dispersion; CSEW = Crime Survey of England and Wales.

[a] MS Excel does not have a formula for range. To calculate range, all that is needed is to take away the minimum figure from the maximum.

The formula above is shown with and without the corresponding data cells that have been used in the CSEW 2016–2017 data set. To include all the data between the cells, all that is needed is to open a bracket, put the first cell that is to be used in the calculation followed by a colon and then the last cell to be used. This will then include all the cells in between. The screen shot in Figure 5.5 shows this in cell 35,422 and also in the formula bar above the data.

Looking ahead

In this chapter, we have introduced the ideas underpinning the use of interval data. As hopefully is clear, this type of data is very different from categorical data and, as such, requires us to develop different types of frequency tables. You would never use MCT to describe a nominal variable and you would never use a frequency table to describe a scale variable. Here, MCT and MoD are used to summarise the data rather

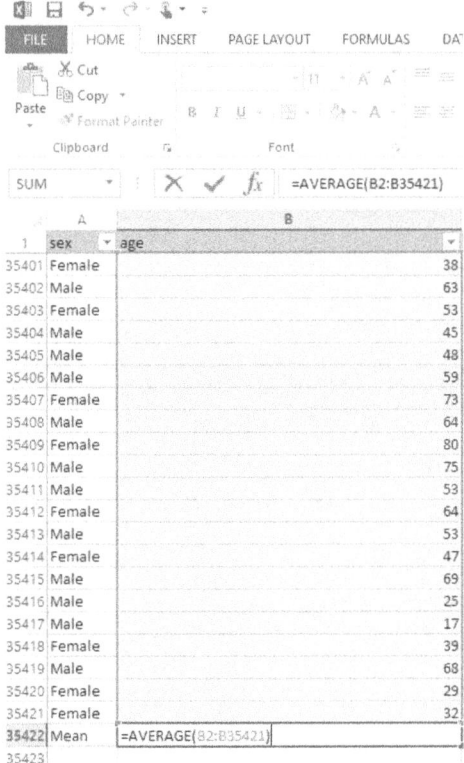

Figure 5.5 MS Excel selecting data from B2:B35421

than valid percent and frequencies. In Chapter 6, we will show how data can be visualised in different ways so that the story it tells can reach a wider audience: visualisations are particularly useful at adding a narrative to scale data, as non-statistically literate people might struggle to interpret some of our MCT/MoD. Tables are useful but can be a little dry. So why not look for different ways to visualise the data in a colourful and interesting way.

Chapter Summary

- When we refer to MCT, we are in fact referring to any of the three averages: the mean, the median and the mode.
- MoD take the data, and rather than summarising which figure represents the middle, they instead produce a summary of the diversity of the data. The simplest MoD to calculate is the range. However, using the range does have its limitations: it's less precise when comparing data from different sources. Standard deviation is a standardised and accurate approach by which the spread of the data is reported.

- Standard deviation can be calculated manually; however, it would be better to let the statistical package of choice do the work.
- When writing up your results, you must be explicit in what is being measured.

Further Reading

Chapter 5 introduced the use (and misuse) of averages. These suggested further readings demonstrate how different authors have used averages to construct interesting statistical narratives.

Charlesworth, S., & Smith, M. (2018). Gender pay equity. In A. Stewart, J. Stanford, & T. Hardy (Eds.), *The wages crisis in Australia: What it is and what to do about it* (pp. 85–102). University of Adelaide Press. www.jstor.org/stable/j.ctvb4bt9f.10

This interesting article explores the gender pay gap that exists in Australia. Again, it presents a powerful narrative using only descriptive statistics, in this instance, averages.

Joseph Rowntree Foundation. (2016). *UK poverty: Causes, costs and solutions.* https://www.jrf.org.uk/report/uk-poverty-causes-costs-and-solutions

While this report by the Joseph Rowntree Foundation is quite long, it is an excellent example of where descriptive statistics are used persuasively and to make a strong political point, in this instance, the use of averages.

Mitchell, K. R., Mercer, C. H., Prah, P., Clifton, S., Tanton, C., Wellings, K., & Copas, A. (2019). Why do men report more opposite-sex sexual partners than women? Analysis of the gender discrepancy in a British National Probability Survey. *Journal of Sex Research, 56*(1), 1–8. https://doi.org/10.1080/00224499.2018.1481193

This is a fascinating journal article that presents a convincing narrative about sex differences when reporting the numbers of sexual partners men and women have throughout their lifetimes. The article often only uses descriptive statistics, in this instance, averages and percentages. Do we finally have the answer for why men over-report and women under-report?

Mercer, C. H., Tanton, C., Prah, P., Erens, B., Sonnenberg, P., Clifton, S., Macdowall, W., Lewis, R., Field, N., Datta, J., Copas, A. J., Phelps, A., Wellings, K., & Johnson, A. M. (2013). Changes in sexual attitudes and lifestyles in Britain through the life course and over time: findings from the National Surveys of Sexual Attitudes and

Lifestyles (Natsal). *Lancet* (London, England)*, 382*(9907), 1781–1794. https://doi. org/10.1016/s0140-6736(13)62035-8

This is another good journal article that uses the Natsal-3 data set to explore changes in sexual attitudes using little more than averages and percentages.

6

VISUALISING OUR DATA

Chapter Overview

Does a picture tell a thousand words?...160

Does it really matter how I present my data? 160

There's a graph for that...164

Spoilt for choice! 189

What makes a good graph?..189

Data mapping 191

Looking ahead..197

Further Reading 198

Does a picture tell a thousand words?

This chapter is all about how to visualise our data. If you have been reading every chapter, especially the past three, you might be forgiven for thinking what were those frequency tables, if not visuals. Well they are visual, although we think you would agree they are not exactly eye-catching. When we talk about **visualising data**, we probably actually should call it graphing data, which doesn't sound very nice but simply means visualising our data using bar, pie, line and all sorts of other charts and graphs that allow us to visualise our data without actual words and numbers. Increasingly, visualisation of data also refers to geospatial or digital data maps and suchlike, which are becoming increasingly common and complex. A good data visualisation can provide a much more compelling and dramatic summary of our data than a long narrative summary; we live in an increasingly visual and busy world, good visualisations often have the greatest impact because they are easy and quick to digest. We should also note that many of us are visual learners and communicators and therefore we may find visualisation of data an easier way to encounter and analyse data. Students often don't appreciate the difference between types of data visualisations but when they get it wrong, they can lose both marks and the impact of their greater data narrative. Whether you have read all the previous chapters or are just dipping into this one, this chapter will introduce a range of types of data visualisations and show you the benefits and limits of each.

Does it really matter how I present my data?

Visualisations of data are often presented as an add-on to complement a frequency table, deemed to somehow enhance it. This is certainly the approach our students usually take and statistical software encourages this because it is so easy to run everything at the same time. We need to reconsider what data visualisations can do for us. If you read Chapter 1, you may recall we mentioned something called 'exploratory data analysis' which was first popularised by someone called Tukey? Well, Tukey (1970) and others proposed that the first stage of any data exploration should be graphical and that we should start with our 'uncleaned' data. Tukey thought that visualisations would allow the researcher to see initial patterns or trends in the data, prior to the researcher manipulating the data through cleaning. Because lots of numbers (or text) easily distract the eye, a visual can force us to really 'see' the data. This approach means that you have to appreciate what sort of function each graph does. Additionally, it would probably be quicker than ploughing through lots of frequency tables or MCT outputs. This approach also avoids the researcher imposing an early manipulation of the data with the potential loss of interesting trends, anomalies or patterns. To illustrate the point, Table 6.1 and Figure 6.1 show the distribution of the variable 'respondent's sex' (*rsex*) across the levels of the variable 'Feel satisfied with sex life' (*sffsatis*). This data is from the Natsal-3 data set (http://www.natsal.ac.uk) that we mentioned in Chapter 2. Compare how easy it is to spot the data trends in Table 6.1 and Figure 6.1. One look at Figure 6.1 and we can see a trend of females being more likely to 'agree' and 'strongly agree' than men, suggesting that females may be more likely to 'strongly agree/agree' that they are satisfied with their sex lives than are men. Of course, we would need to do more exploration of the data but just by running the bar chart we can see a potential trend to explore. Table 6.1 features the same data and with a little time, you would identify the same trend.

Table 6.1 Cross-tabulation of rsex and sffsatis

Feel satisfied with sex life * Respondent's sex Crosstabulation

| | | | Respondent's sex | | |
			Male	Female	Total
Feel satisfied with sex life	Agree strongly	Count	1361	1994	3355
		% within Respondent's sex	22.9%	23.7%	23.4%
	Agree	Count	2280	3209	5489
		% within Respondent's sex	38.4%	38.1%	38.2%
	Neither agree or disagree	Count	1244	2070	3314
		% within Respondent's sex	21.0%	24.6%	23.1%
	Disagree	Count	824	855	1679
		% within Respondent's sex	13.9%	10.1%	11.7%
	Disagree strongly	Count	224	300	524
		% within Respondent's sex	3.8%	3.6%	3.6%
Total		Count	5933	8428	14361
		% within Respondent's sex	100.0%	100.0%	100.0%

Just to further emphasise Tukey's point about spotting patterns in the data, have a look at Figure 6.2, which is a scatterplot featuring data again from Natsal-3.

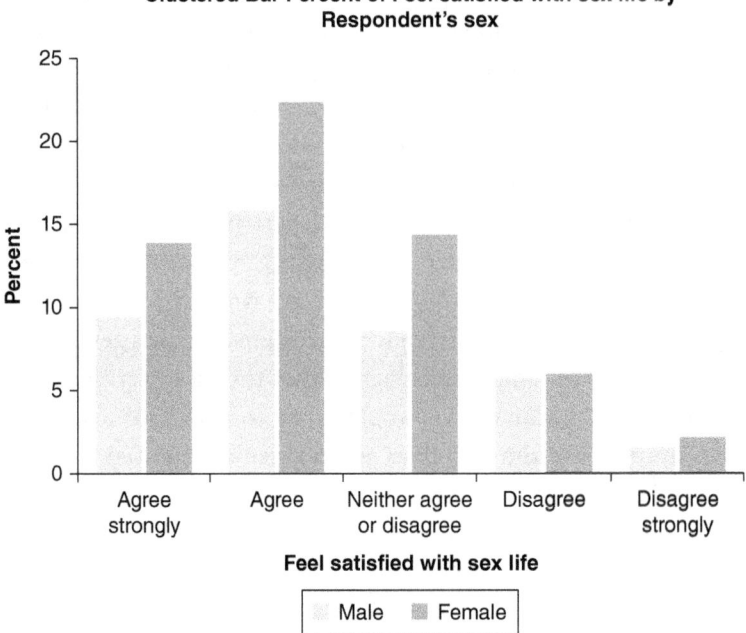

Figure 6.1   Clustered bar chart of rsex and sffsatis

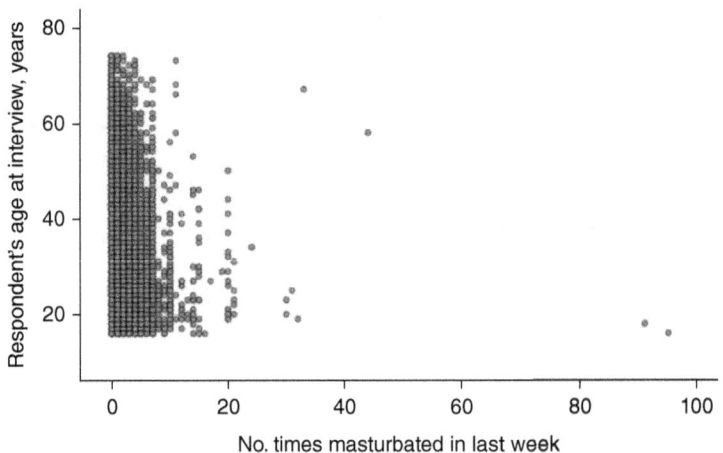

Figure 6.2   Scatterplot of dage and mastnum

Figure 6.2 explores the possible relationship between age and 'number of times a respondent masturbated in a week'. We can see that there are some potential outliers, most of whom appear to be younger people, and we can see that the trend in the data is for level of masturbation to remain constant irrespective of age. If we were to examine MCT for this variable, it would take much longer to come to this initial conclusion. By eyeballing Figure 6.2 we can see some potentially interesting trends but also consider that other factors may drive masturbatory levels, irrespective of age. Hopefully, you appreciate Tukey's point about using graphs and other visuals to initially encounter and explore your data.

The other central role that visualisations of data can play is in communicating our findings, our 'story'; after all, they say a picture tells a thousand words. In communicating our narrative, graphs can do a number of things:

- Present the data clearly to the reader in an easy to digest manner
- Challenge the reader to think about the 'story' that the data tells
- Present often large and complex statistics, simply
- Make complex data and analysis simple and coherent
- Communicate the big impactful 'message' of your research

It should also be noted that in this era of short attention spans and digital communication, a really great visual can make a big impact in a way a wordy summary might not; you will probably find that good visualisation of your data is an important research and employability skill. The good news is that the majority of statistical software have graph and other visualisation tools to make it easier for you to visualise your data. The quality of these tools does vary and there are online graphics packages that you might want to use to enhance your visuals. This chapter uses the IBM SPSS chart builder tool to generate its graphs (and other outputs). This chapter does not tell you how to run specific visuals on specific software, but it does identify the key ways to visualise your data. Of course, you can also plot graphs manually, which is fine for small data sets but may be a challenge too far for larger ones. There are lots of great guides online as to how to manually plot different types of graphs.

Box 6.1

Pause for Thought

Reflect on the following questions:

- Do you have a usual type of graph that you always use and why?
- How much time have you spent in the past creating visuals of your data?
- Are graphs central to your work or always an 'add-on'?
- How might data visualisations assist your future data analysis?

There's a graph for that

It can seem a bit bewildering at first when confronted by all the different types of graphs that can be used to visualise data, particularly as they may seem very similar at first. This section will not cover every single type of graph, but it will focus on the main types and discuss when each is most useful. Many decisions about which graph to choose relate to the type of data that you have and what the graph aims to achieve. If the graph is there as a means to preliminarily explore your 'raw' data then you may choose a simple scatterplot or bar chart; if you are using the graph to tell a powerful story of your data then an alternative graph, such as a cluster bar or a line graph, may be more appropriate. There are two central rules to apply to all graphs:

1   Always ensure that everything is labelled correctly and clearly.
2   Favour simplicity over complexity in design.

The one with the slices

One of the very first graphs you may have come across at school is the pie chart, so called because of its resemblance to a pie, with each slice representing a specific category or level of a variable, either as a count or as a percentage (or both). Pie charts are a good way to visualise nominal or ordinal data and can communicate a clearer story than a table; compare Table 6.2 on page 165 and Figure 6.3 below, which show data from the GSS (2018, http://GSS 2018.norc.org), which was discussed in Chapter 2.

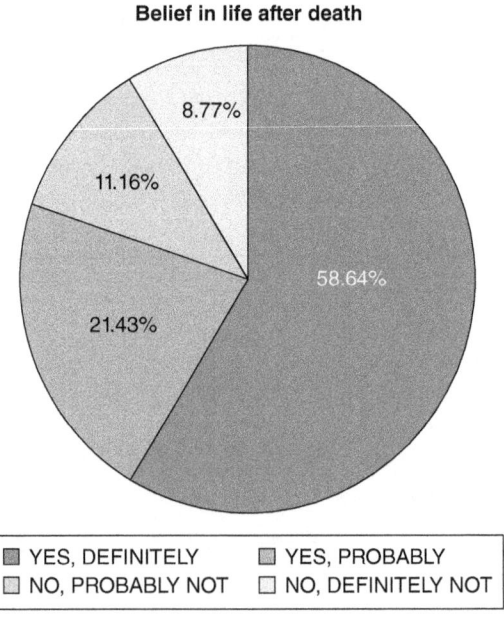

Belief in life after death

8.77%
11.16%
21.43%
58.64%

☒ YES, DEFINITELY   ☒ YES, PROBABLY
☐ NO, PROBABLY NOT   ☐ NO, DEFINITELY NOT

Figure 6.3   Pie chart for AFTERLIF

When we eyeball Figure 6.3 we can see that the majority (58.64%) of GSS 2018 respondents who answered the question about 'belief in life after death' said 'yes, definitely'. Table 6.2 also tells us this, but I bet you identified the key trend quicker looking at the pie chart. You should never use pie charts to visualise scale data, if you want to know why then look at Figure 6.4, which shows respondent's age from the GSS 2018; but don't look at it for too long as you may get dizzy!

Table 6.2 Frequency table for AFTERLIF, GSS 2018

Belief in life after death

		Frequency	Percent	Valid Percent	Cumulative Percent
Valid	YES, DEFINITELY	662	28.2	58.6	58.6
	YES, PROBABLY	242	10.3	21.4	80.1
	NO, PROBABLY NOT	126	5.4	11.2	91.2
	NO, DEFINITELY NOT	99	4.2	8.8	100.0
	Total	1129	48.1	100.0	
Missing	IAP	1173	50.0		
	DK	36	1.5		
	NA	10	.4		
	Total	1219	51.9		
Total		2348	100.0		

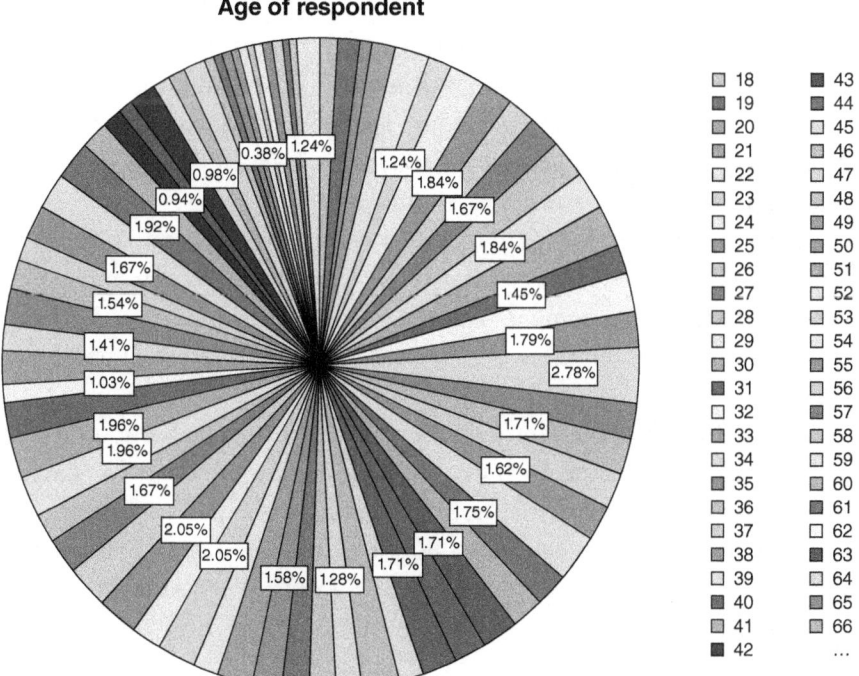

Age of respondent

Figure 6.4 Pie chart for AGE, GSS 2018

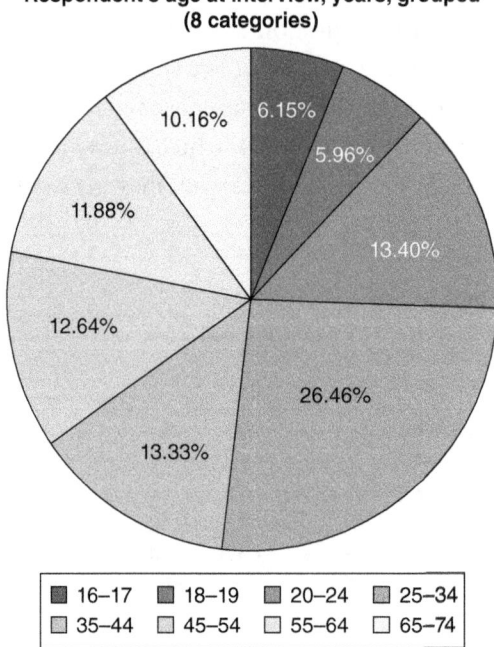

Respondent's age at interview, years, grouped (8 categories)

6.15%
5.96%
13.40%
26.46%
13.33%
12.64%
11.88%
10.16%

| ■ 16–17 | ■ 18–19 | ■ 20–24 | ■ 25–34 |
| ■ 35–44 | □ 45–54 | □ 55–64 | □ 65–74 |

Figure 6.5   Pie chart for dagegr, from Natsal-3

Pie charts work best when they display a variable with less than six categories; the more categories the more difficult it is to identify the specific slices of the pie, as Figure 6.5 demonstrates. Figure 6.5 shows eight age group categories from Natsal-3. You can still interpret Figure 6.5 (the largest age group is 25- to 34-year-olds), but it would probably take you longer than eyeballing a frequency table, as it is quite a 'busy' pie chart with several categories of similar size.

A common pie chart mistake is to forget to insert the percentage or count within the actual slice, you need this information so that the reader can understand the proportions. To illustrate, compare Figure 6.6 and Figure 6.7. Figure 6.6 is colourful, but it doesn't tell us anything except to identify the different categories within the data.

Florence Nightingale, the famous 19th century British nursing pioneer, popularised the pie chart in the 19th century. She used a pie chart to analyse patterns of mortality amongst British soliders during the Crimean war, noting that the majority of deaths were caused by poor hygiene and had a seasonal element to them. Her pie chart (reproduced in Figure 6.8) easily communicated her message to the authorities back in the UK, prompting action to improve military mortality rates.

**Respondent's age at interview, years, grouped
(6 categories)**

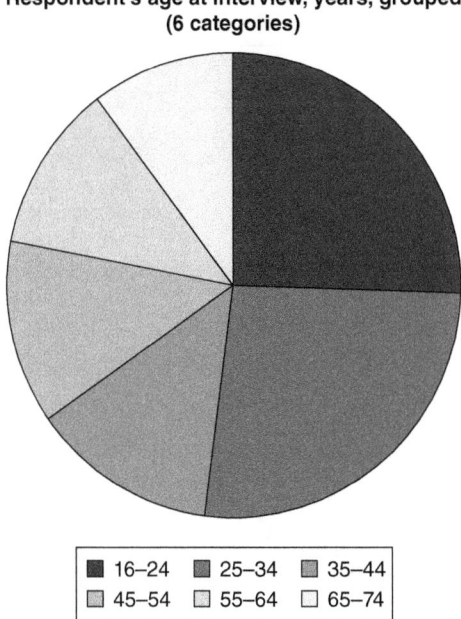

| ■ 16–24 | ■ 25–34 | ■ 35–44 |
| □ 45–54 | □ 55–64 | □ 65–74 |

Figure 6.6   Pie chart for agrp, from Natsal-3

**Respondent's age at interview, years, grouped
(6 categories)**

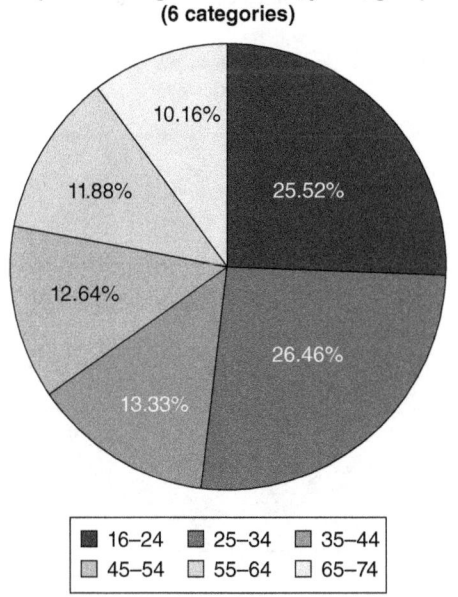

| ■ 16–24 | ■ 25–34 | ■ 35–44 |
| □ 45–54 | □ 55–64 | □ 65–74 |

Figure 6.7   Pie chart for agrp, from Natsal-3

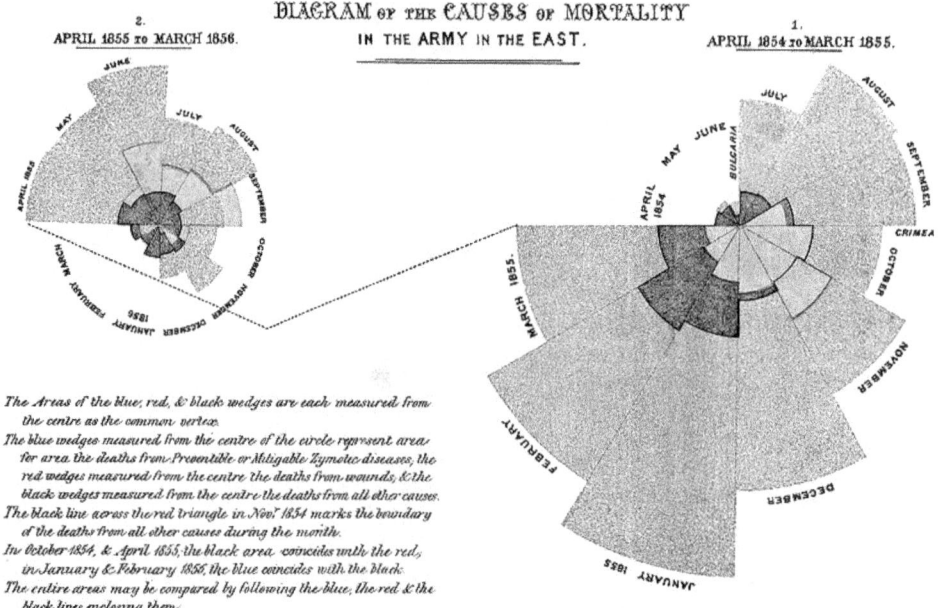

Figure 6.8 Mortality amongst British soldiers during the Crimean war

Source. https://commons.wikimedia.org/wiki/File:Nightingale-mortality.jpg

Figure 6.8 shows how the mid grey areas are preventable deaths caused by poor hygiene. The light grey areas are deaths from wounds and the black areas are all other deaths. It is clear that the majority of deaths are preventable.

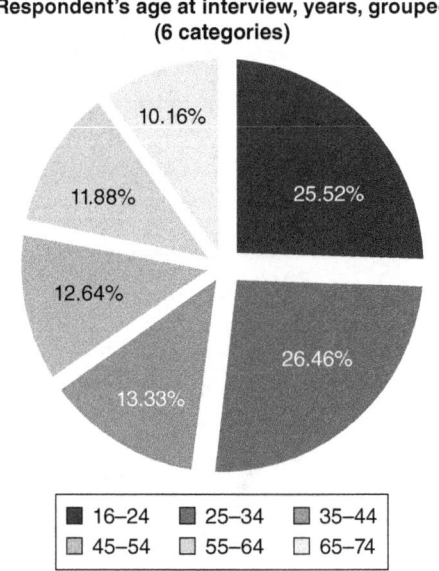

Figure 6.9 'Exploded' pie chart for agrp, from Natsal-3

There are many variations of the standard pie chart, including 3D (three-dimensional) and 'exploded' pie charts; Figures 6.9 and 6.10 demonstrate each of these. The point of adopting different styles of pie chart is to create an eye-catching visual to assist your reader in following your narrative.

Respondent's age at interview, years, grouped (6 categories)

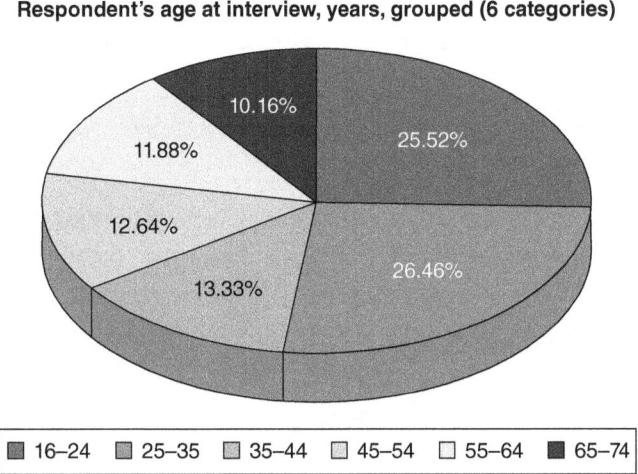

| ■ 16–24 | ■ 25–35 | □ 35–44 | □ 45–54 | □ 55–64 | ■ 65–74 |

Figure 6.10   '3D' pie chart for agrp, from Natsal-3

The one with the bars

Another graph that you probably learnt at school was the bar chart; please note that you can call these 'bar charts' or 'bar graphs', as it refers to the same thing, the important thing is to be consistent, so pick one and use it throughout your work. Bar charts are used to visualise and/or compare percentage, frequency or a measure (e.g. the mean or the median) for the distinct categories within a variable. Bar charts, like pie charts, are amongst the most commonly used means to visualise data because they are easy to interpret. Bar charts should be used for nominal and/or ordinal data but not scale. It is good practice when constructing a bar chart using nominal data to arrange the bars sequentially from the largest to the smallest, or vice versa; this makes it easier for the reader to interpret the data. With ordinal data, you should not arrange the bars sequentially as the different categories have a predetermined sequence. Figure 6.11 shows a simple bar chart featuring nominal data from the GSS 2018 (in this case asking respondents about their *'confidence in the US military'*).

Figure 6.11   Simple bar chart showing nominal data (CONARMY, GSS 2018)

From Figure 6.11 we can clearly see that the majority of respondents have a 'great deal of confidence in the military'. You could change the colour of each bar to add greater contrast to each response. Figure 6.12 shows a simple bar chart featuring ordinal data: in this case, GSS 2018 respondents were asked '*Life serves no purpose?*'

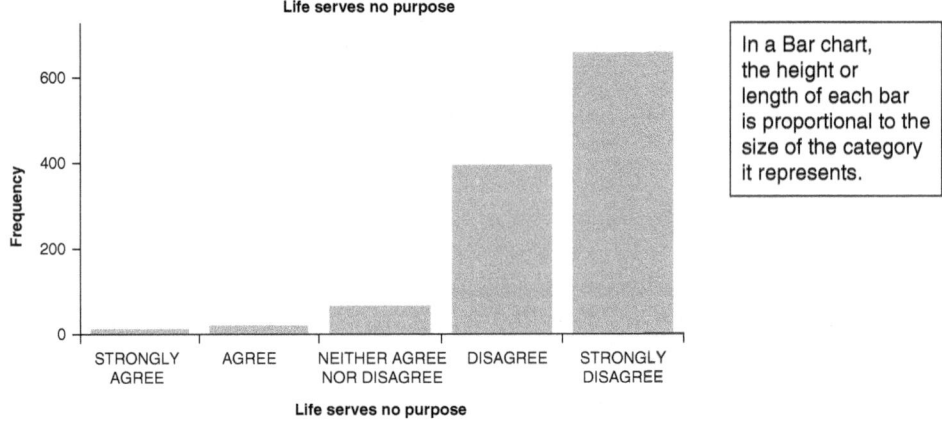

Figure 6.12   Simple bar chart showing ordinal data (NIHILISM, GSS 2018)

Did you spot the majority response in Figure 6.12? The majority clearly 'strongly disagreed/disagreed' with the view that '*Life serves no purpose*'.

There are many different types of bar chart, the one you choose depends on the story that you want your graph to communicate. Figure 6.13 shows a horizontal bar chart showing data from the GSS 2018, specifically a variable *AFRAIDOF* that asks the respondent if 'People act as if they are afraid of R'.

We interpret horizontal bar charts in much the same way as vertical ones; the longer the bar, the larger the category. Horizontal bar charts can be useful when you have long data labels or titles, which can be difficult to fit below the *x*-axis. Likewise, if we

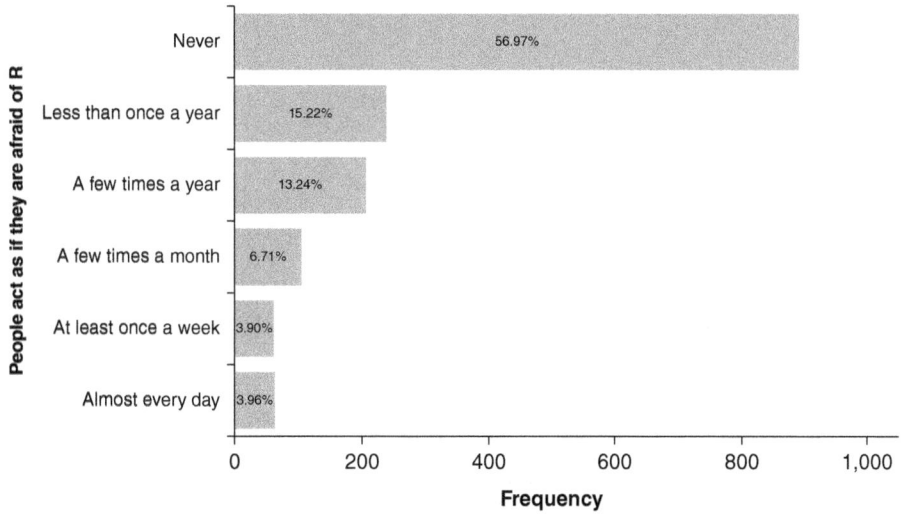

Figure 6.13   Horizontal bar chart showing ordinal data (AFRAIDOF, GSS 2018)

have many different categories within our variable, it can be more clearly presented as a horizontal bar chart. If we arrange our categories in order of the largest first (as shown in Figure 6.13), it can have more impact if presented as a horizontal rather than as a vertical bar chart. Finally, if we have negative values, then it can be more easily represented in a horizontal bar chart. You will have noted that Figure 6.13 has the percentage values for each category added to provide the reader with additional information; this is something you may decide is useful to add to your visual or you may think that it merely clutters it. You may also think that so far, a simple frequency table might have been just as easy to interpret, well you may be right but what if you had a cross-tabulation. Table 6.3 shows a cross-tabulation of data from the GSS 2018: respondent's gender (*GENDER1*) and 'respondent feels that family make too many

Table 6.3   Cross-tabulation of GENDER1 and DEMANDS (GSS 2018)

R feels that family make too many demands on R * Gender of 1st person Crosstabulation

Count

		Gender of 1st person		Total
		MALE	FEMALE	
R feels that family make too many demands on R	No, never	370	326	696
	Yes, but seldom	131	97	228
	Yes, sometimes	85	92	177
	Yes, often	24	20	44
	Yes, very often	13	11	24
Total		623	546	1169

demands on R' (*DEMANDS*). Compare Table 6.3 with Figure 6.14, which visualises the data using a vertical bar chart.

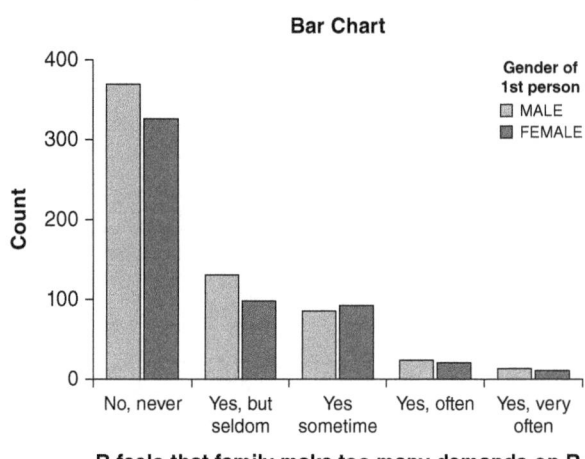

Figure 6.14 Grouped bar chart of DEMANDS and GENDER1 (GSS 2018)

We wonder which one you found easier to interpret? In addition, did you spot the trend; there does not seem to be a great difference between male and female responses to this question.

A grouped or clustered bar chart is one that features more than one grouping variable. In Figure 6.14, we have the distribution of responses to the question grouped by the two categories of gender. In a simple bar chart, we would only have the responses to the question generally but not its grouping by distribution across the levels of another variable. In this type of bar chart, we can easily compare responses to the question by males and females. Figure 6.14 shows what is technically known as a double bar chart or graph. You can technically have as many subgroupings as you want, but if you have too many, your graph will become increasingly cluttered: remember to keep things simple.

Stacked bar charts are another way to visualise your data. Figure 6.15 shows both a vertical and a horizontal stacked bar chart, featuring the variables, *RELPERSN* (which asks whether the respondent 'considers self a religious person') and *MARHOMO* (which asks whether 'Homosexuals should have right to marry') from GSS 2018.

Stacked bar charts are similar to grouped/clustered bar charts except that the different coloured bars are stacked on top of each other as opposed to being side by side. Each colour represents a different subgroup or, in this case, response on a Likert scale of the *MARHOMO* variable. Stacked bar charts are good for showing the total size of

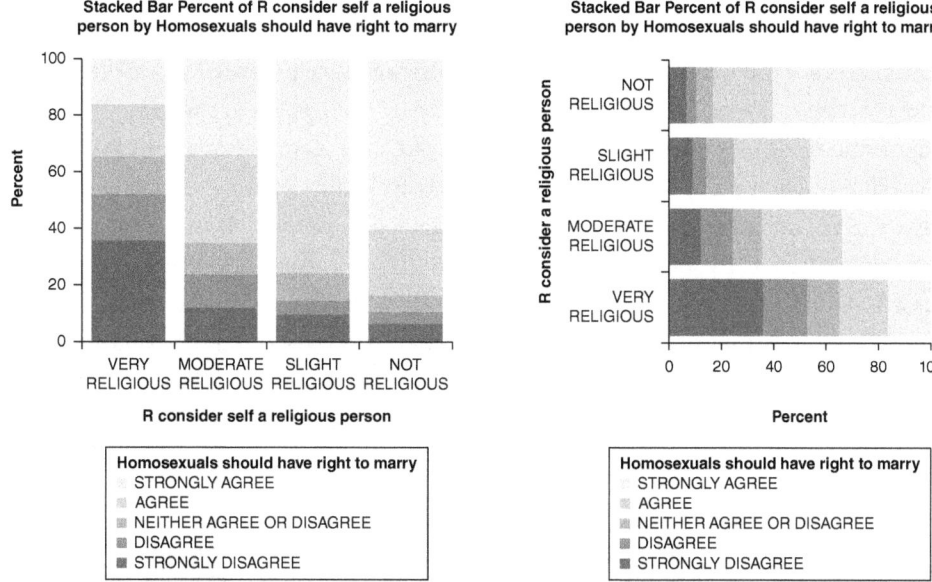

Figure 6.15 Vertical and horizontal stacked bar charts showing ordinal data (RELPERSN and MARHOMO, GSS 2018)

groupings; thus, if we look at all the responses for each category of religiousness, we can see what percentage out of the total responded per answer to the variable *MARHOMO*. Thus, we can get a sense of proportionality across categories and responses. If we look at Figure 6.15, we can see that around 50% of the responses of 'very religious' people are 'strongly disagree/disagree' to the question of homosexuals having the right to marry; in contrast over 80% of the responses of 'not religious' people are 'strongly agree/agree' to the same question. Stacked bar charts can be vertical or horizontal, although some consider the horizontal orientation to be more eye-catching and therefore impactful. Again, we can contrast the visual with a cross-tabulation of the two variables; compare Table 6.4 with Figure 6.15, which do you find easier to interpret?

Table 6.4 Cross-tabulation of RELPERSN and MARHOMO

Homosexuals should have right to marry * R consider self a religious person Crosstabulation

Count

		R consider self a religious person				
		VERY RELIGIOUS	MODRTE RELIGIOUS	SLIGHT RELIGIOUS	NOT RELIGIOUS	Total
Homosexuals should have right to marry	STRONGLY AGREE	39	197	168	212	616
	AGREE	45	181	107	82	415
	NEITHER AGREE NOR DISAGREE	32	66	36	21	155
	DISAGREE	41	71	18	15	145
	STRONGLY DISAGREE	87	68	34	22	211
Total		244	583	363	352	1542

The one with the lines

Line charts are the same as bar charts except instead of bars they use lines; please note that you can call these 'line charts' or 'line graphs', as it refers to the same thing, the important thing is to be consistent, so pick one and use it throughout your work. Although the two types of charts are similar, typically line charts are used to visualise trends over time; if your data is not showing such a trend then use a bar chart. To illustrate, Figure 6.16 shows data from the GSS 2018 (using the *GSS Explorer tool* discussed in Chapter 2) represented by a simple line chart showing the 'always wrong' responses to the question 'is it wrong for same-sex adults to have sexual relations' (*HOMOSEX*) over the period 1973–2018. We can clearly see that the trend is downward, suggesting a liberalisation of attitudes towards homosexuals.

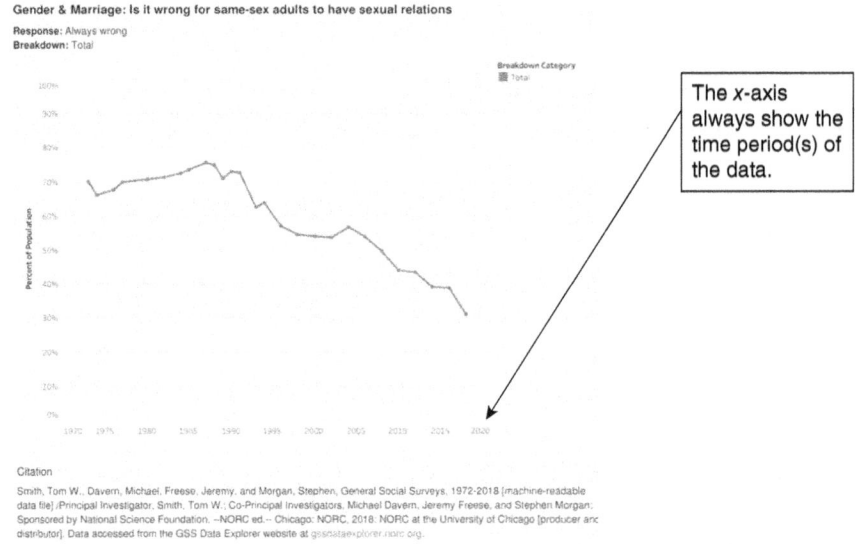

Figure 6.16 Simple line chart of HOMOSEX, GSS 1973–2018

A bar chart could also show this data, but the impact of the visual would be lessened. Likewise, avoid using line charts to show anything but temporal trends, as Figure 6.17 demonstrates: the line chart is confusing and hard to interpret, compare it with Figure 6.12, we are sure you will agree that Figure 6.12 is a much clearer way of showing the trend in the data.

Simple line charts visualise one variable over time, as Figure 6.16 demonstrates. However, we might want to explore the difference between categories or levels of a variable over time. Figure 6.18 visualises the differences over time between men and women's attitudes to the variable *HOMOSEX*. The trend is again downward for both but the female line is lower than the male one and drops more sharply, suggesting that female views tend more toward the liberal in relation to this question.

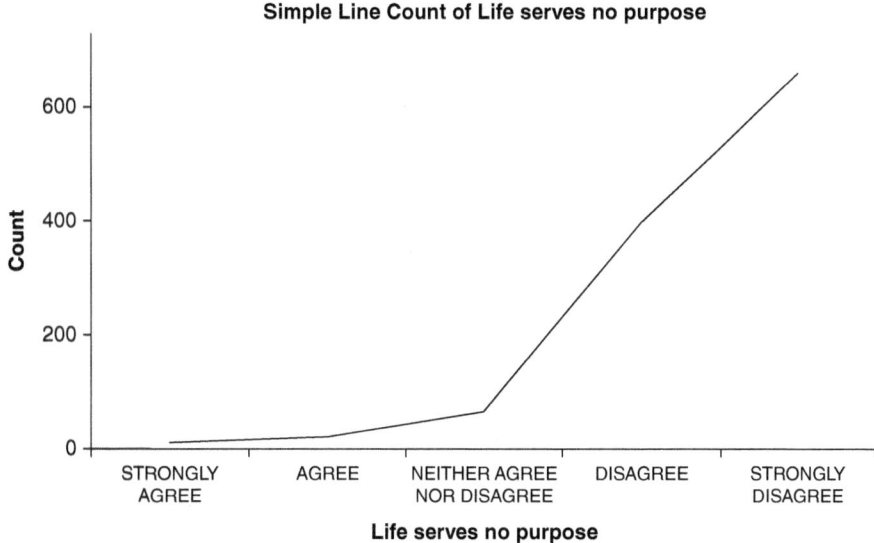

Figure 6.17 Simple line chart of NIHILISM, GSS 2018

Citation

Smith, Tom W., Davern, Michael, Freese, Jeremy, and Morgan, Stephen, General Social Surveys, 1972-2018 [machine-readable data file] /Principal Investigator, Smith, Tom W.; Co-Principal Investigators, Michael Davern, Jeremy Freese, and Stephen Morgan; Sponsored by National Science Foundation. --NORC ed.-- Chicago: NORC, 2018: NORC at the University of Chicago [producer and distributor]. Data accessed from the GSS Data Explorer website at gssdataexplorer.norc.org.

Figure 6.18 Multiple line chart of HOMOSEX and SEX, GSS 1973–2018

Figure 6.19 shows another multiple line chart, this time showing age groups and the *HOMOSEX* variable.

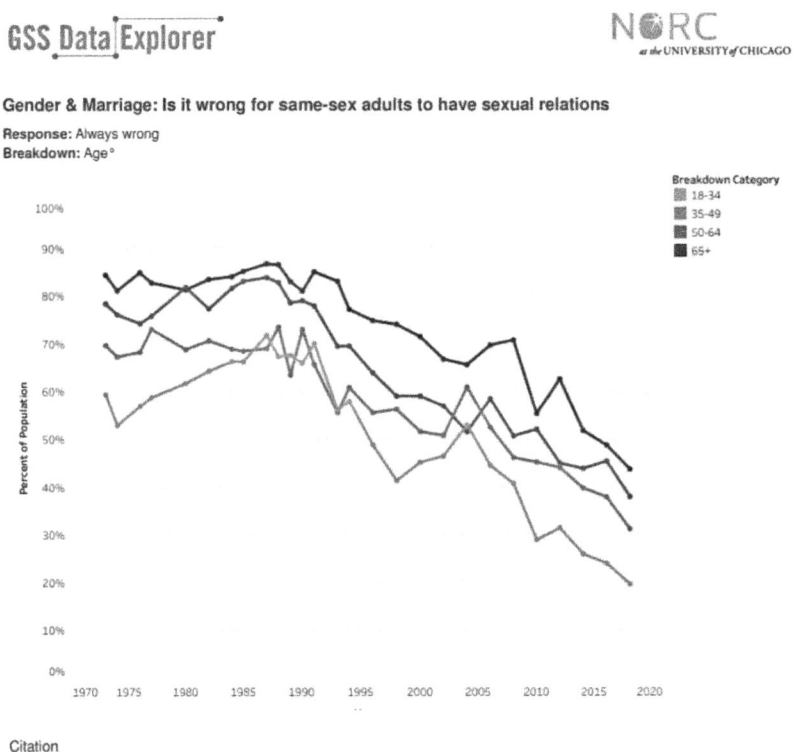

GSS Data Explorer

NORC at the UNIVERSITY of CHICAGO

Gender & Marriage: Is it wrong for same-sex adults to have sexual relations

Response: Always wrong
Breakdown: Age°

Breakdown Category
18-34
35-49
50-64
65+

Citation

Smith, Tom W., Davern, Michael, Freese, Jeremy, and Morgan, Stephen, General Social Surveys, 1972-2018 [machine-readable data file] /Principal Investigator, Smith, Tom W.; Co-Principal Investigators, Michael Davern, Jeremy Freese, and Stephen Morgan; Sponsored by National Science Foundation. --NORC ed.-- Chicago: NORC, 2018: NORC at the University of Chicago [producer and distributor]. Data accessed from the GSS Data Explorer website at gssdataexplorer.norc.org.

Figure 6.19 Multiple line chart of HOMOSEX and AGE, GSS 1973–2018

Box 6.2

Reflective Exercise

Look at Figure 6.19 and try to summarise the key trends. Then go to GSS Data explorer and generate some simple line graphs using other variables of your choice.

The one with that sounds like instagram

So far, we have been discussing ways to visualise nominal and ordinal data but you are probably wondering about scale data. A common way to visualise scale data is by using a histogram. It is a graph or plot that allows you to visualise and examine the frequency distribution or shape of a scale variable. We can identify a range of useful information about our data including type of distribution, skewness and outliers. Let's start by looking at a standard table (Table 6.5) of MCT for the variable, *periods2*, which asks the respondent at what 'age when started menstruating', taken from Natsal-3. It features the usual array of statistics that we would expect to see; can you spot the trend in the data?

Table 6.5 MCT for periods2 from Natsal-3

Statistics

Age when started menstruating

N	Valid	8774
	Missing	6388
Mean		12.89
Median		13.00
Mode		13
Std. Deviation		1.581
Variance		2.501
Range		19
Minimum		7
Maximum		26
Percentiles	25	12.00
	50	13.00
	75	14.00

Figure 6.20 shows what is known as a simple histogram; a simple histogram shows the frequency of one variable: Did you spot the trend in the data quicker by eyeballing Figure 6.20?

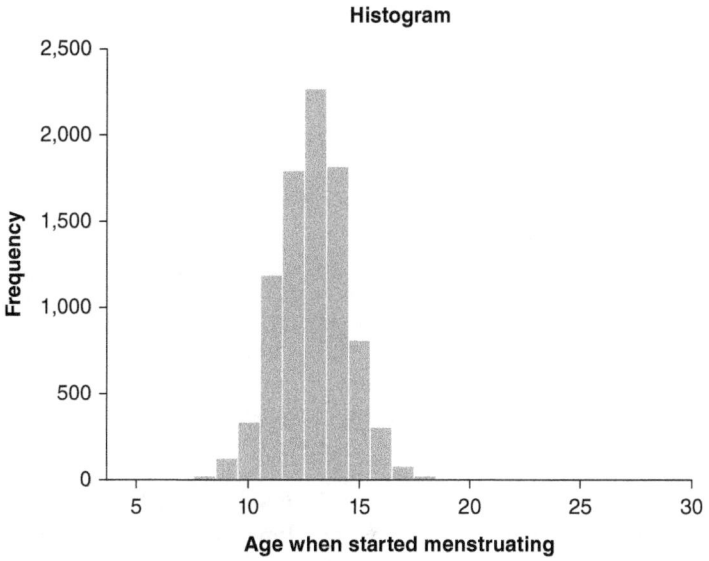

Figure 6.20 Simple histogram of periods2 from Natsal-3

Histograms are a great way to communicate trends in scale data without overwhelming the reader (or yourself) with lots of MCT statistics.

On our histogram (Figure 6.20), the x-axis shows our ages and the y-axis shows the actual frequencies per age. The histogram splits the data into intervals known as bins. In Figure 6.20, each bin is 1 year of age, but if we had a larger variable, then we might change our interval. We can also alter the width of each band to make our visual clearer. Histograms, unlike bar charts, do not have gaps between each bar because the data featured is scale and not categorical. If we look at Figure 6.21, we can see that it looks like an approximately normal distribution with the greatest number of respondents having started menstruating between 12 and 14 years, it doesn't appear to have any skew, but it is a narrow and tall distribution, suggesting a narrow range of scores. There do not seem to be any outliers. In many software packages, we can add the line of normal distribution to our histogram to aid interpretation as shown in Figure 6.21. When we add the distribution line, it appears to reinforce our view that this data is approximately, normally distributed; although we should obviously further interrogate the data through the running of MCT and MoD.

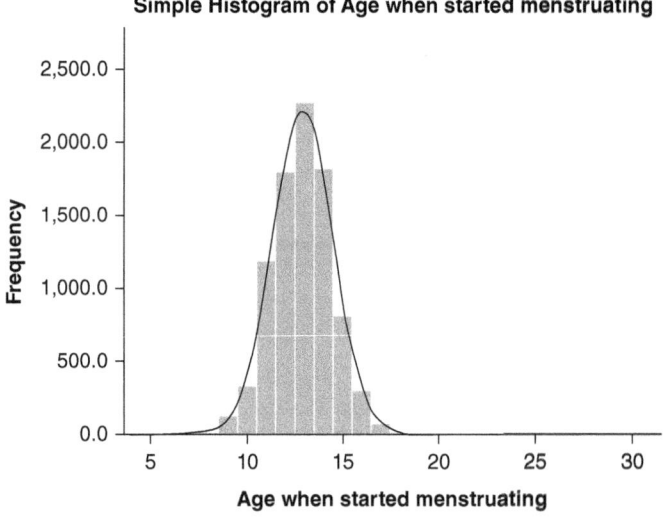

Figure 6.21 Simple histogram of periods2 from Natsal-3, with line of distribution included

There may be times when we want to visualise the distribution of a scale variable across the levels or groups of a categorical variable. We can do this by running what is known as a stacked histogram as shown in Figure 6.22. It features two variables from Natsal-3: *dage* (respondent's age when they completed the survey) and *contrameth* (which asks 'whether the respondent used a reliable contraceptive for their first sex').

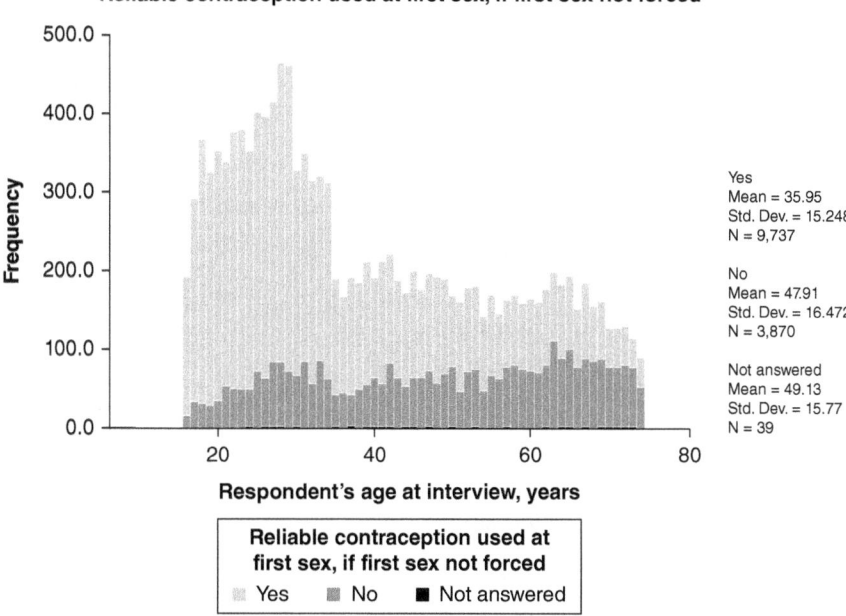

Figure 6.22   Stacked histogram of dage and contrameth from Natsal-3

We can see that the trend in the data suggests a relationship between age and contraception use at first sex: the older the person the less likely they were to have used
contraception at first sex, alternatively we could say that the younger the person the
more likely they were to use contraception at first sex. It would be much harder to
identify the trend if we had a large table of MCT like the one shown in Table 6.6.

Table 6.6   MCT of dage and contrameth from Natsal-3

Descriptives			
Respondent's age at interview, years	**Reliable contraception used at first sex, if first sex not forced**		**Statistic**
	Yes	Mean	35.95
		95% Confidence Interval for mean	Lower bound	35.65
			Upper bound	36.26
		5% Trimmed mean	35.15
		Median	31.00
		Variance	232.507
		Standard deviation	15.248
		Minimum	16
		Maximum	74
		Range	58

(Continued)

Table 6.6 (Continued)

Respondent's age at interview, years	Descriptives			Statistic
	Reliable contraception used at first sex, if first sex not forced			
	No	Mean		47.91
		95% Confidence Interval for mean	Lower bound	47.39
			Upper bound	48.43
		5% Trimmed mean		48.13
		Median		49.00
		Variance		271.315
		Standard deviation		16.472
		Minimum		16
		Maximum		74
		Range		58
		Minimum		24
		Maximum		73
		Range		49

You would, eventually, work out the trend in the data from Table 6.6, but we bet you would identify it more quickly through looking at Figure 6.22. There are two other types of histograms that you may want to use. One is as a frequency polygon (see Figure 6.23) and is the same as a histogram except that it uses a line to show the frequency distribution and the area below the line is shaded.

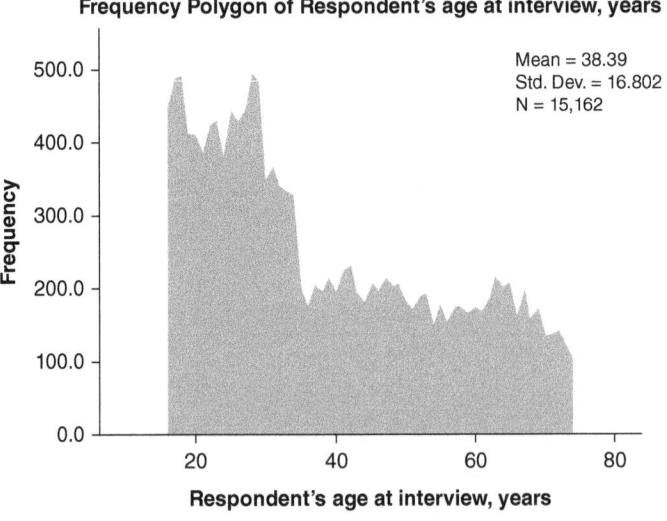

Frequency Polygon of Respondent's age at interview, years

Mean = 38.39
Std. Dev. = 16.802
N = 15,162

Figure 6.23 Frequency polygon of dage from Natsal-3

Figure 6.24 shows the same data as a simple histogram; they obviously look pretty similar and which one you choose is really up to you and sometimes it may come down to a design decision. You may also want to try visualising your data using a population pyramid (see Figure 6.25).

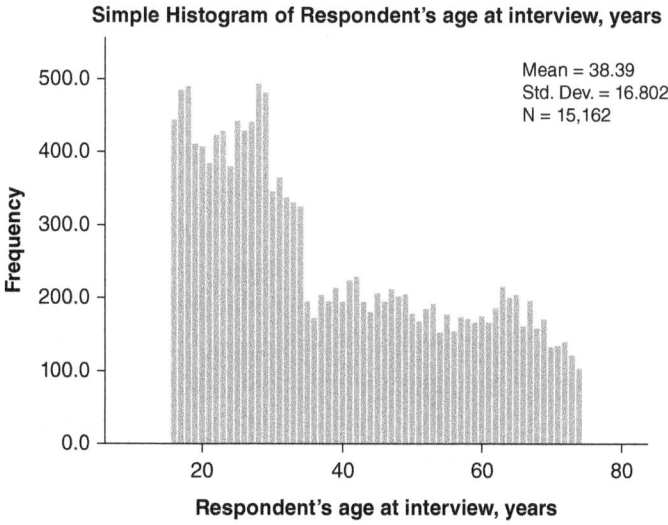

Figure 6.24 Histogram of dage from Natsal-3

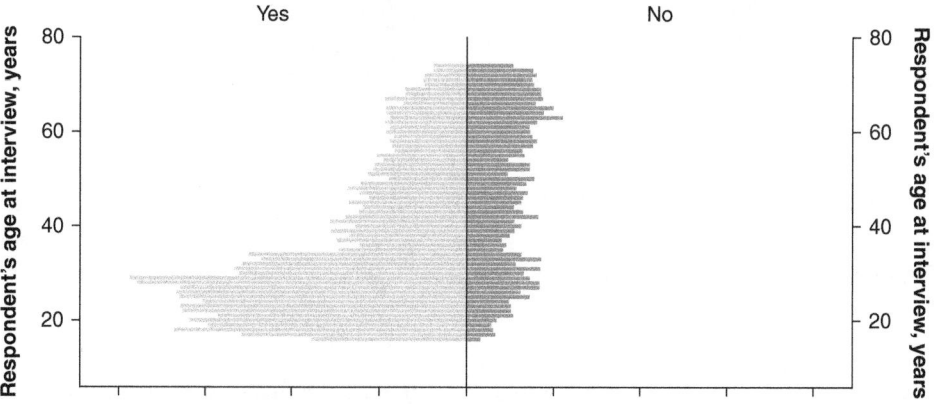

Figure 6.25 Population pyramid of dage and contrameth from Natsal-3

Figure 6.25 shows the same data that we used in our stacked histogram, but this time the variable identifying age is on the vertical axis and the frequencies are on the horizontal axis. The two categories of the *contrameth* variable ('yes' and 'no') are back to back on the graph. Population pyramids are really good for comparing distributions across groups, in the case of Figure 6.25, we can clearly see that overall the majority did use contraception the first time they had sex but that the trend relating to the 'no' group is for older people to not have used contraception.

The one with the whiskers

Another graph that you can use to visualise and explore scale data are boxplots, which are also known as box and whisker diagrams/plots. Again it does not matter which title you use, as long as you are consistent in your usage of it. Boxplots show distributions of data but lack the detail of a histogram, but they do present some key information about data, especially about the distribution, its central value and variability; they are also useful for identifying potential outliers. Boxplots work really well with large data sets and are perfect for comparing multiple distributions. Figure 6.26 shows the simplest type of boxplot, the 1D Boxplot, of the variable *AGE* from GSS 2018.

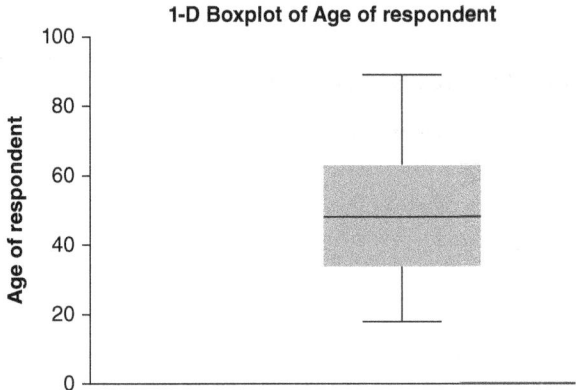

Figure 6.26   1D boxplot of AGE, GSS 2018

Compare the ID boxplot shown in Figure 6.26 with the histogram of the same data shown in Figure 6.27, which do you find easiest to interpret?

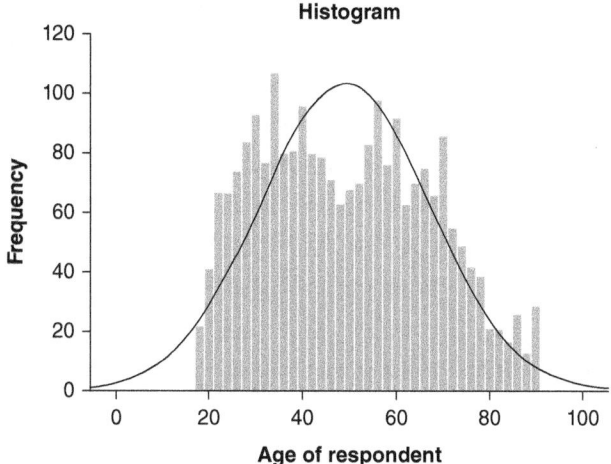

Figure 6.27 Histogram of AGE, GSS 2018

Boxplots are great because they present key summary information about the data (in this example the distribution of the variable age), as Figure 6.28 shows.

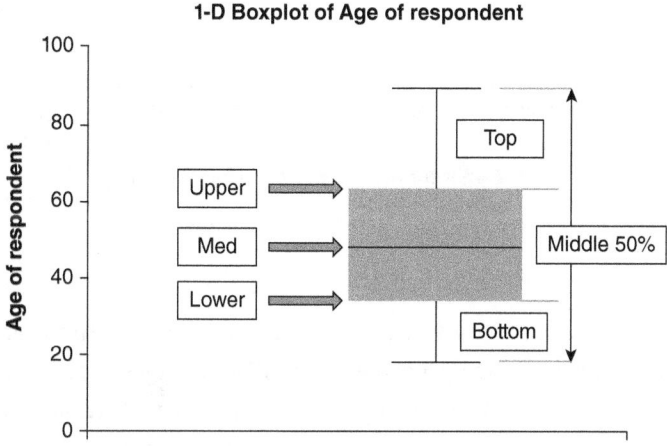

Figure 6.28 Guide to 1D boxplot of AGE, GSS 2018

The line in the middle of the box is the median; the ends of the box are the upper and lower quartiles, thus the box spans the interquartile range (IQR); the 'whiskers' are the two lines that extend to the highest and the lowest observations or data points. For our Age variable, we can see from Figure 6.28 that the median is around 48, and that around 50% of our distribution is between 20 and 60 years. There are no outliers, but the youngest respondent is 19 and the oldest is 90. The top 'whisker' is longer than the bottom one, which shows that there are more ages spread out in the upper quartile compared with the bottom quartile. Figure 6.27 shows the distribution of the same data but visualised in a simple histogram.

Boxplots are particularly useful when we want to compare distributions, such as in Figure 6.29 which shows the distribution of age across the different categories or levels of the *FEFAM* variable from the GSS 2018, which asks respondents 'is it better for man to work, woman tend home'. This sort of boxplot is known as a simple boxplot.

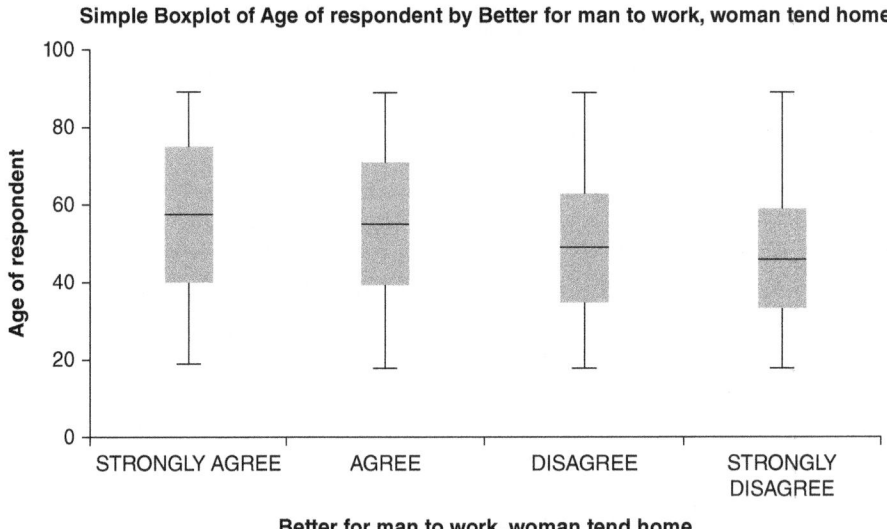

Figure 6.29 Simple boxplot of Age and FEFAM, GSS 2018

When we eyeball Figure 6.29, we can instantly see that there are differences between the distribution of age per category of *FEFAM*, suggesting an age difference in views regarding gender roles. For example, 50% of the age distribution for 'strongly disagree' is between 60 and 35, with the lowest median age, whereas the 'strongly agree' distribution has the highest median and 50% of the respondents fall between 40 and 75. This data suggests that older people tend towards the 'strongly agree/ agree' categories that reinforce traditional gender roles. The simple boxplot is probably easier to interpret than a stacked histogram.

We can also create clustered boxplots, which allows us to review the distribution of our scale variable in relation to two other variables, further splitting the data. If we take our example from Figure 6.29 and add gender into the mix, we get Figure 6.30, note that the second variable is in a different colour to make interpretation easier.

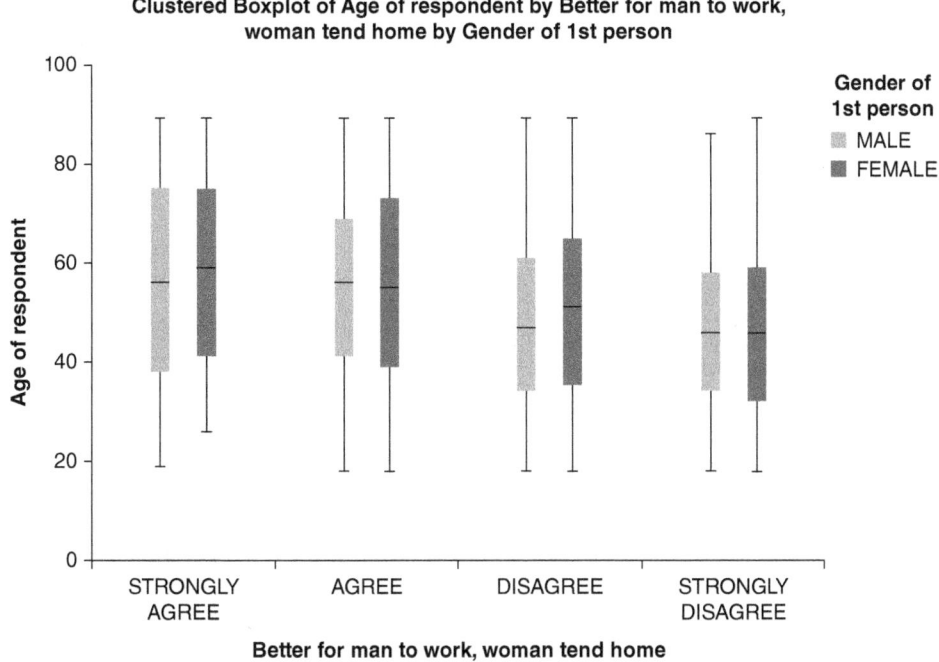

Figure 6.30 Clustered boxplot of Age, FEFAM and Gender, GSS 2018

Figure 6.30 suggested an age difference in terms of respondents' views on traditional gender roles. When we add gender into the analysis, things change a bit. For the men, the median reduces across the levels, between 'strongly disagree' to 'strongly agree', suggesting that the younger a man is the more likely he is to have non-traditional views on gender roles. This is reinforced by the IQR of each male group; in the negative categories, it is wider but tending towards the upper age ranges, whereas the positive categories have shorter IQRs, tending towards the younger ages. For women, the median ages are higher than men for the 'strongly agree' and the 'disagree' categories and the IQR are wider than men's in all categories bar 'strongly agree'. This gender difference may suggest that for women, their gender may be more important than their age in determining their views on traditional gender roles, whereas for men, age may be the more important factor. Of course, this data would need much greater examination before we could 'draw' clear conclusions about the trends in the data.

The one with all the dots

The final graph that we will review is the scatterplot. Scatterplots allow us to visualise and examine the relationship between variables as opposed to their frequencies. A scatterplot is a graph that plots a respondent's score on one variable against their score on a second variable, thus, showing a potential relationship between the two variables. The scatterplot has two coordinates (each representing a variable), the X-coordinate, which is plotted on the *x*-axis (the horizontal axis, thus left or right), and the Y-coordinate, which is plotted on the *y*-axis (the vertical axis, thus, up or down). The dot in the scatterplot represents the intersection of the two coordinates. Scatterplots can tell us four important things about our two variables:

1   Is there a linear or non-linear relationship?
2   Is the relationship positive or negative?
3   Is the relationship strong, moderate or weak?
4   Are there any outliers?

Figure 6.31 shows a simple scatterplot of data from the GSS 2018, specifically respondent's age (*AGE*) and weight (*WEIGHT*); we might presume that these two variables are in a relationship with each other, people often joke about gaining weight as they get older.

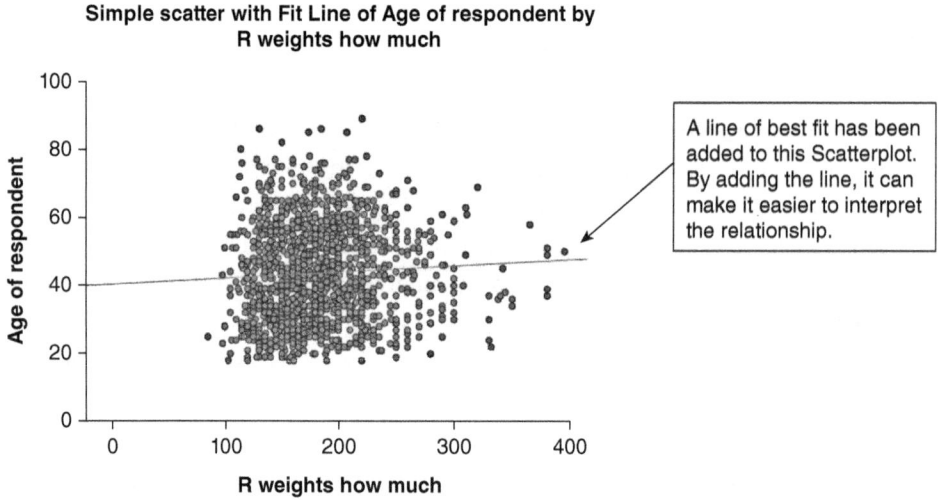

Figure 6.31   Simple scatterplot, AGE and WEIGHT, GSS 2018

Figure 6.31 suggests that there is a positive linear relationship between age and weight, in other words as your age increases so does your weight. That said, the line is not steep, indeed it is quite flat suggesting the relationship is weak. There are also, what appear to be, many outliers in the upper weight scores which may be having an effect. If we wanted to further explore the relationship between age and weight, we could add a categorical variable, for example gender, to see how being male or female influenced the relationship between age and weight. To do this, we can create a grouped scatterplot, which uses colour to distinguish between the different groups of our categorical variable, as shown in Figure 6.32.

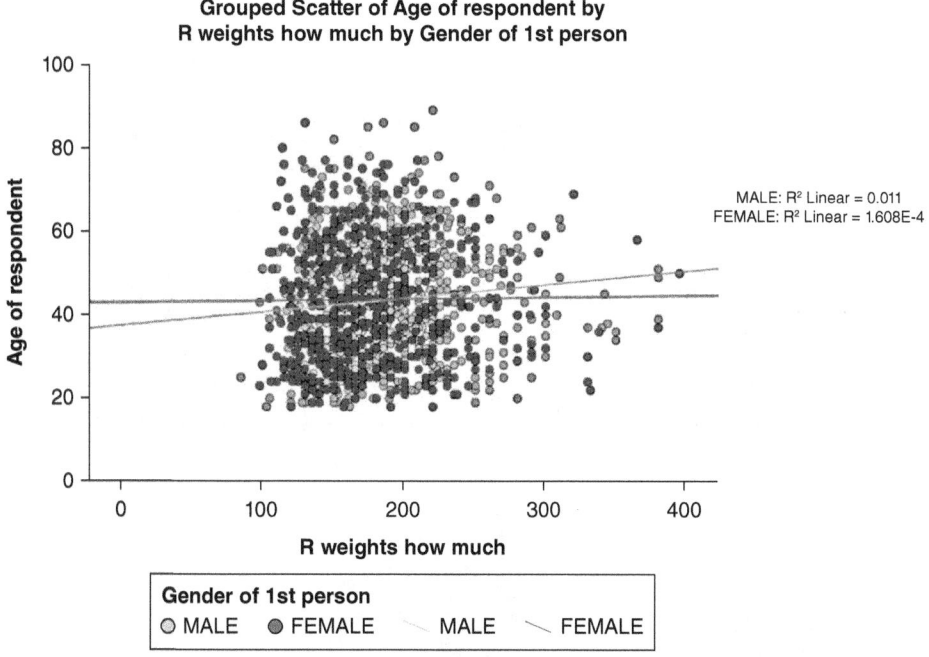

Figure 6.32 Grouped scatterplot, AGE, WEIGHT and GENDER

You will note that we have added two lines of fit, one for each gender, this time; the male line is steeper than the female (the female line is almost flat) suggesting that the relationship between age and weight is stronger for males and positive, there also appears to be a lot of outliers which may be skewing the data. One final way you might want to explore your data is by using a matrix scatterplot, which allows you to explore the relationships between three variables. Figure 6.33 shows data from the GSS 2018, specifically 'age of respondent' (*AGE*), 'hours per day spent watching TV' (*TVHOURS*) and 'number of hours worked last week' (*HRS1*). We might think that the younger you are, the more you work and so the less time you have to watch television; conversely, we might think that the older you are the less you work and the more time you might have to watch television. A matrix scatterplot is six scatterplots in one visualisation. The three scatterplots above the diagonal are the same as the ones below the diagonal but with their axes reversed.

You will notice that for our matrix scatterplot we have added lines of best fit for each scatterplot. We can see that age and hours of TV watched per day is a positive

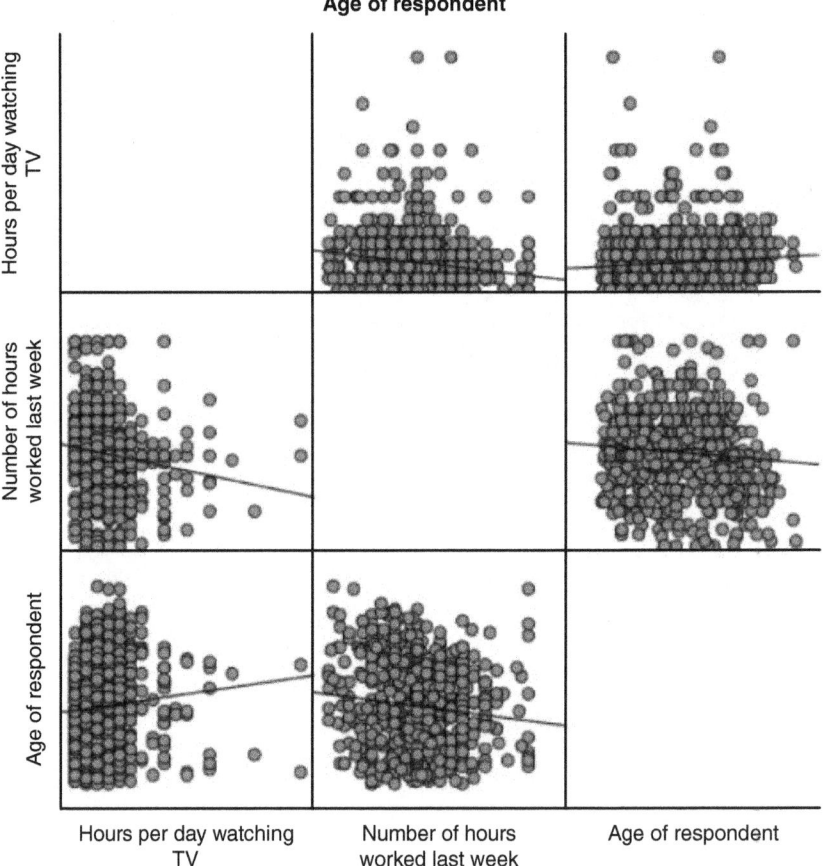

Figure 6.33 Matrix scatterplot, AGE, HRS1 and TVHOURS, GSS 2018

and strong relationship, as we get older, we watch more television, but we should also note that there are clear outliers who are watching a lot of television. Age and hours worked is in a negative relationship, which looks quite strong; as we get older, we work less hours, which is good news! Finally, bad news for TV fans, the more we work, the less television we watch per day, well at least until we retire!

Box 6.3

2-Minute Recap!

Set your timer and see how many types of graphs you can list in two minutes. Make it competitive by competing against a classmate(s).

Spoilt for choice!

So, we have looked at a lot of different graphs and charts, some let us make comparisons, others let us identify potential relationships. Some are appropriate for categorical data, others for scale only. You may be feeling a bit overwhelmed with information and choice. This may be amplified by the fact that some graphs may do the same sort of thing as others, like bar and line charts; then there is the issue of different varieties of the same sort of chart. Remember some golden rules. Firstly, make sure you are using the right sort of graph for your data, for example, no pie charts featuring scale data or histograms featuring categorical data. Secondly, think about what you want your visual to do for you – Is it to explore data or create a narrative impact in an assessed submission? Finally, what is the absolutely best 'fit' for the story that you are trying to tell your reader when she or he reads your research report?

What makes a good graph?

The previous section highlighted an array of different graphs and charts that you may want to use to visualise your data. Apart from deciding which graph is most suited for your data, you also need to ensure that your graph 'works' from a design point of view. It is easy to spot a bad graph that is one that doesn't help the reader to 'see' the data but rather merely confuses. Can you spot the problems with the two graphs – Figures 6.34 and 6.35?

At first glance, Figure 6.34 looks okay but it has a classic pie chart error – the percentages add up to more than 100! Watch out for this, as it is a very common error.

Figure 6.35 is a classic error that you will see in the media; the baseline of 0% has been omitted which makes the bar sizes look much larger than they actually are, which of course is misleading. The omission of baselines or the absence of an axis/mislabelling of an axis is a widespread means that is used to manipulate data; such graphs are known as truncated graphs. Figure 6.36 shows the same bar chart but this time with a baseline on the y-axis.

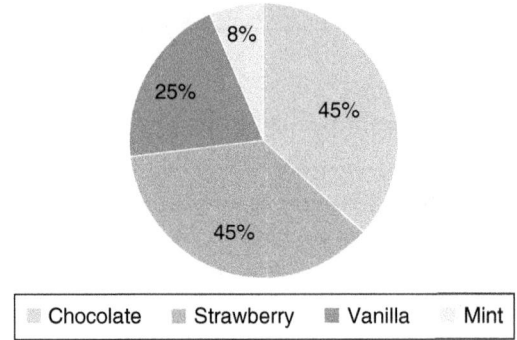

Figure 6.34 Pie chart showing student ice cream preference

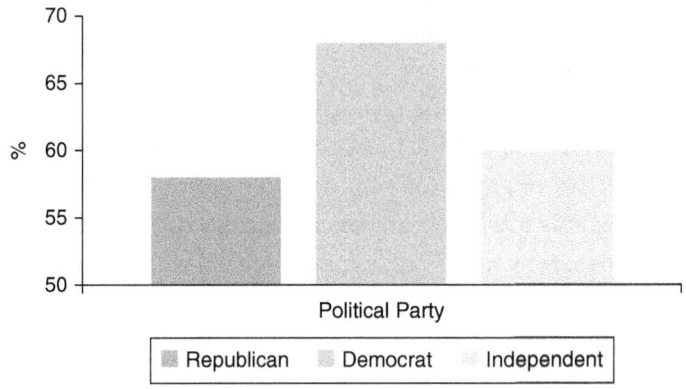

Figure 6.35 Bar chart showing political party preference – Presidential Election

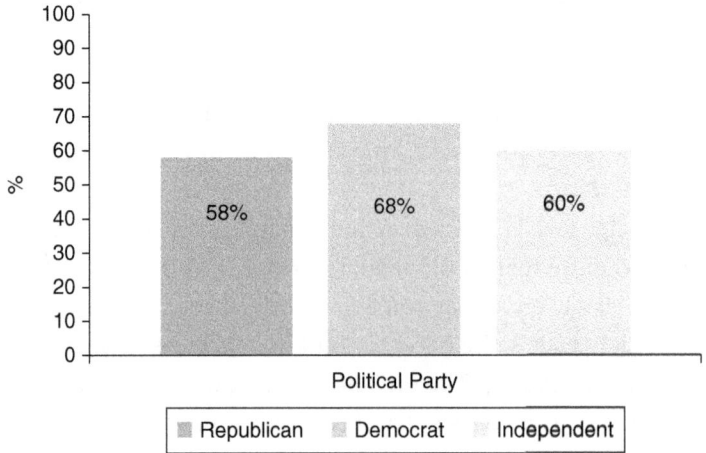

Figure 6.36 Bar chart (with baseline on y-axis) showing political party preference – Presidential Election

There is still a difference between the voter preferences but it appears less than in Figure 6.35 graph. Of course, you won't be setting out to manipulate your graphs in this way, but it is good to be aware of how commonly the media, governments and businesses deliberately use poor visuals to make (or hide) often controversial points; we need to get better at calling them out.

Box 6.5

Time to Get Your Hands Dirty!

Go online and try to find as many really bad or misleading graphs as you can; screen-shot them and save them. With some classmates, share your findings and see if they can spot the errors/problems with each visual. You could make this competitive by seeing how many each of you can gather in a set time.

Always remember the following basic design rules:

- Label your graph clearly and accurately, including axes.
- Use colours that are attractive and impactful but not so bright that they dazzle the reader.
- Have a clear index to the chart so that the reader understands what each colour, bar, slice, line and so on represents and how it is measured – that is, is it the percentage or the count.
- Give your graph a title either above it or below it so that the reader can identify what it is.
- Consider the size of the chart, if it is too small with lots of detail then it will be hard to read, but if it is too large with little detail it will look like you are 'padding' out your work.
- Don't make the graph 'too busy' with lots of detail; if this is your graph then maybe you need to change it to a graph that can accommodate the data better.
- Ensure that your numbers add up, this is particularly often an error found in pie charts.

Remember that if you violate these design rules then your visual will make your work hard to interpret, confuse the reader and/or cause the reader to misinterpret your results.

Data mapping

There is another source of data visualisation that you may want to use as a means to explore trends and patterns – and that is data maps. Data maps typically make use of what is known as geospatial data. This is data that includes objects, events and/ or specific phenomena that have a geographical location. The location may be static (e.g. an earthquake, the location of a specific address, a high crime neighbourhood) or dynamic (e.g. a moving object like a person, car, infectious disease, etc.). Geospatial data combines and links different data together to create vivid maps that visualise the data, often in real time. Geospatial data makes use of Geographic Information Systems, which are different methods for collecting and analysing spatial and geographic data. If you have ever used Google Earth™ or Google Street View™ (https:// earth.google.com/) then you have used Geographic Information Systems data. The growth of more powerful software and increasingly complex databases has led to the growth of geospatial data. However, its origins lie in pre-computer days when researchers used pen, paper and a lot of walking to make their own data maps.

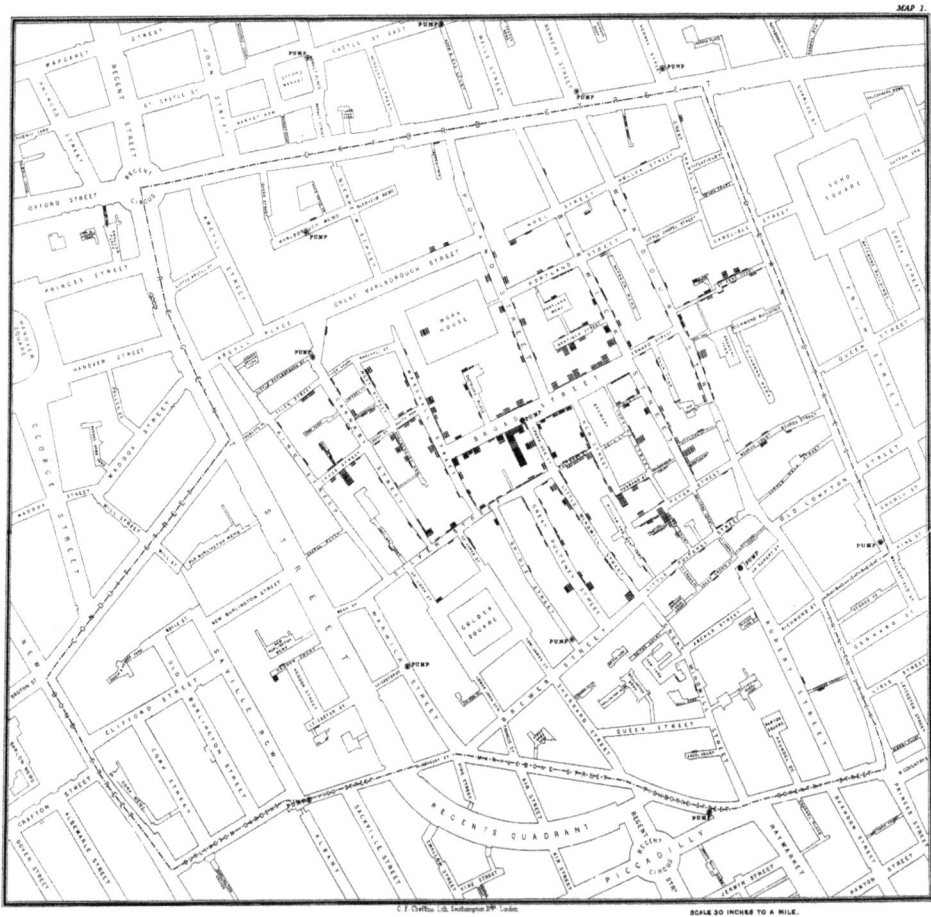

Figure 6.37 Dr John Snow's Cholera map of Soho, London

Source. https://commons.wikimedia.org/wiki/File:Snow-cholera-map-1.jpg

A great example of this is the Victorian medical researcher Dr John Snow. In the early 1850s, London was rocked by a series of cholera epidemics, in an age before an understanding of microbiology and the link between infection and poor sanitation. Doctors believed that the cholera was caused by 'bad air', but Snow was sceptical because he had noticed that there appeared to be clusters of deaths in specific areas (typically poor areas); if it had been 'bad air' then surely the distribution of the disease would have been wider and more random. Snow decided to go to one of the high cholera areas of London and investigate. He walked around the area, marking on a map a bar (see Figure 6.37) for every cholera death, he talked to the locals and gathered information on who had died. He created what we would know today as a data or geospatial map (see Figure 6.38). When he reviewed his map, he noticed that the deaths clustered around a water standpipe where the water was contaminated with sewage (this is before London had proper sewers); he disabled the standpipe and the cholera epidemic ended. Although Snow didn't know what was in the water (i.e. bacteria) that caused the disease, he did understand that the water and sewage were somehow the source of the problem.

Figure 6.38 Detail of Dr John Snow's Cholera map of Soho, London

Source. https://commons.wikimedia.org/wiki/File:Snow-cholera-map-1.jpg

Just imagine what Snow could have achieved with today's computer technology! Today we use geospatial data to understand complex social phenomena and trends. Crime mapping, for example, is increasingly being used by governments to understand crime better and therefore target policing (and other) responses more effectively. This book isn't about geospatial data, indeed that would take an entire book on its own; but it is worth you knowing that there are a great many online sources of such data that you could use and that are free to access. To illustrate, let's have a look at the UK Police's crime maps tool, which is extremely user-friendly and interesting.

You can access UK Police crime maps via their website (https://www.police.uk/); then you click on the 'Find Your Neighbourhood' button on the front page as shown in Figure 6.39.

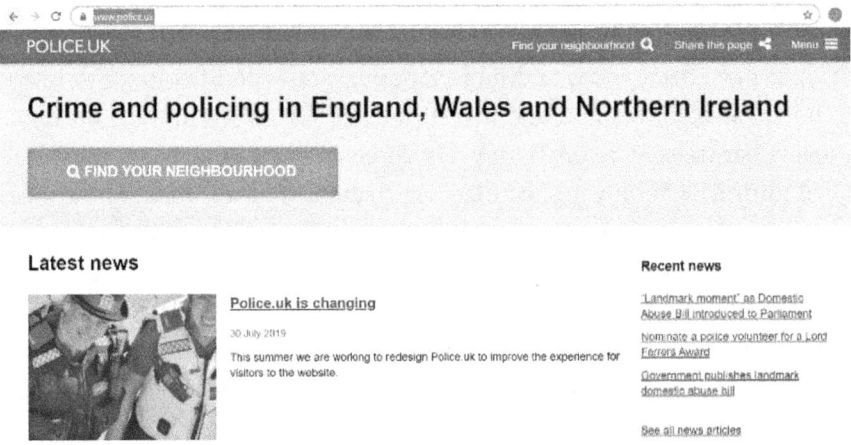

Figure 6.39 'Find Your Neighbourhood' on Police UK website

It prompts you to insert a location or postcode from the UK and then it takes you to a crime map for that area. In our example (Figure 6.40), we used the location 'University area, Manchester'. We can see in Figure 6.40, the entire university area in Manchester with three large crime clusters; you can then zoom in and click on one of the clusters and take a closer look.

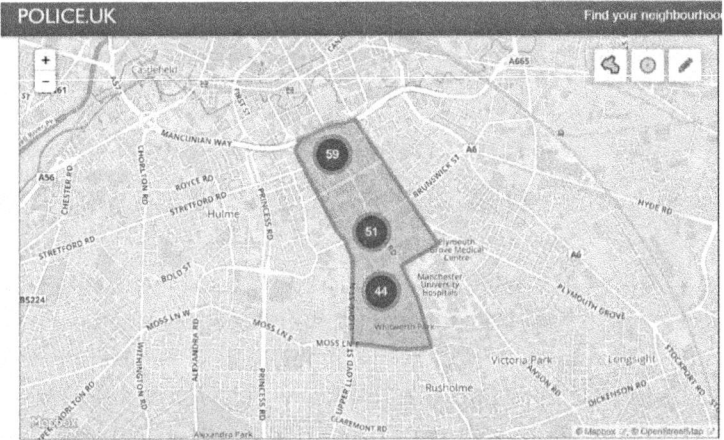

Figure 6.40 Crime map of the University district, Manchester, Police UK website

Let's click on the '59' which takes us to a more detailed map, as shown in Figure 6.41.

Figure 6.41 'Sub-area' of Crime map of the University district, Manchester, Police UK website

You can then zoom in further and hover over one of the numbered circles, which shows you a list of the crimes that occurred (see Figure 6.42) and you can click for further information.

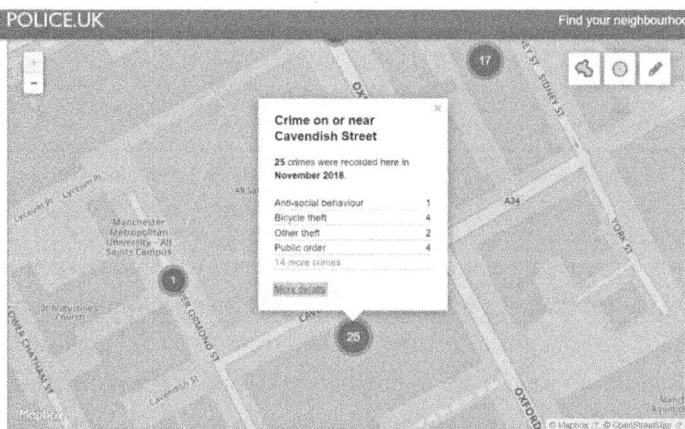

Figure 6.42 List of crimes, sub-area of Crime map of the University district, Manchester, Police UK website

You can see how useful this sort of mapping might be to the local police service in terms of being able to identify clusters of types of crimes and then implement an intervention. To illustrate, the majority of crimes in the university area are antisocial behaviour, bicycle theft and public order. Therefore, the police can do campaigns around bicycle security and work with campus security to ensure that sources of anti-social behaviour are addressed on campus. You could use this data to analyse crime patterns within specific neighbourhoods and then link that to data (e.g. Census data) on poverty, low educational attainment and so forth. Another example of a great source of geospatial data is DataShine (https://datashine.org.uk/) which links geo-spatial data to UK Census data. You can search a specific location and then explore different variables visually. Figure 6.43 shows Manchester's spread of population who have higher education qualifications (18.8% of the population), the areas that are white have no residents with such qualifications, the areas that are darker in colour have higher concentrations. It will not surprise you to learn that the white coloured areas have high levels of socio-economic deprivation. You can download PDFs of your chosen maps, which could be inserted into a research report.

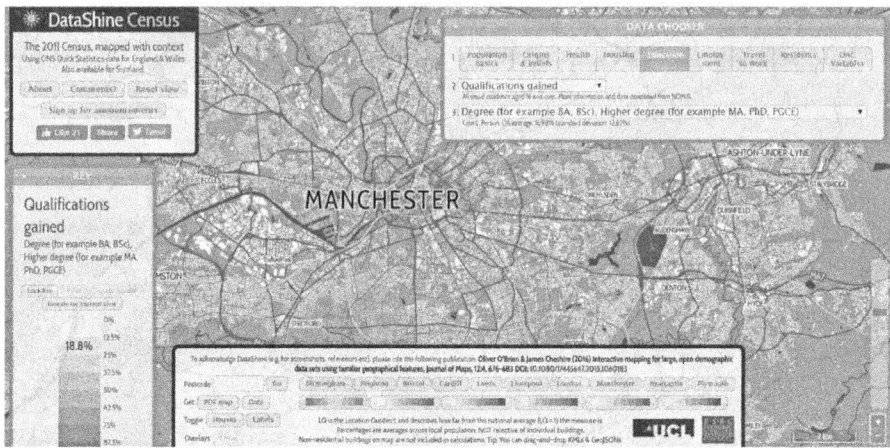

Figure 6.43   Higher Education Qualifications map, Manchester, DataShine

These are just two examples and if you go online, you will find many more. You could use these to explore data, to try and think of a 'story' that you want to pursue; you could use such maps to do your actual analysis either directly or by linking to a data set; or you might use visuals from such maps to enhance your research report. These sorts of data maps are transforming how we understand complex social phenomena, like crime or educational inequalities; maybe a picture really does tell a thousand words!

Box 6.6

Time to Get Your Hands Dirty!

Go online and locate an open access source of geospatial data, like DataShine. Take some time to explore the data: Can you develop a potential data story for the map?

Think about how the creators of the source have linked different data sources together: Can you identify some of them?

Can you think of how this data might inform a service, system, product or policy change?

Looking ahead

This chapter talked a lot about storytelling; how you can use visuals to enhance stories and indeed to tell the whole story of your data. That leads us to our next and final chapter where we focus on the art of telling interesting and important stories with our data. All the previous chapters have been about specific elements of data collection, management and analysis. However, we never collect data for the sake of it; we collect data because we want to do two important things. Firstly, we want to identify or uncover something interesting about social phenomena, whether a new trend, changing pattern or a persistent 'wicked' problem. Secondly, we want to tell people (other social researchers, tutors, policymakers, the government and so forth) about what we have found and why we think it is important. This book has shown you how you do the first thing, and in the final chapter, we will show you how to do the second.

Chapter Summary

- Visualising data simply means visualising data using bar, pie, line and all sorts of other charts and graphs that allow us to visualise our data without actual words and numbers.
- We live in an increasingly visual and busy world, good visualisations often have the greatest impact because they are easy and quick to digest. In this era of short attention spans and digital communication, a really great visual can make a big impact in a way a wordy summary might not.
- Graphs are useful in presenting the data clearly, challenging the reader. The majority of statistical software have graph and other visualisation tools to make it easier for you to visualise your data.
- If the graph is there as a means to preliminary explore your 'raw' data, then you may choose a simple scatterplot or bar chart; if you are using the graph to tell a powerful story of your data, then an alternative graph, such as a cluster bar or a line graph, may be more appropriate.
- There are two central rules to apply to all graphs:
  1   Always ensure that everything is labelled correctly and clearly.
  2   Favour simplicity over complexity in design.
- Pie charts are a good way to visualise nominal or ordinal data and can communicate a clearer story than a table.
- A histogram is a graph or a plot that allows you to visualise and examine the frequency distribution or shape of a scale variable. Boxplots show distributions of data but lack the detail of a histogram, but they do present some key information about data, especially about the distribution, its central value and variability; they are also useful

(Continued)

for identifying potential outliers. Boxplots work really well with large data sets and are perfect for comparing multiple distributions.

- Scatterplots allow us to visualise and examine the relationship between variables as opposed to their frequencies.
- There is another source of data visualisation that you may want to use as a means to explore trends and patterns – and that is data map.
- Data maps can also be used to explore trends and patterns and typically make use of what is known as geospatial data. This is data that includes objects, events and/or specific phenomena that have a geographical location.

Further Reading

Frew, S. (2009). *Now you see it*. Analytics Press.

Stephen Frew has written many books on visualising data and the dos and don'ts of visualisation. In this book, he focuses on keeping it simple and how simple visualisation techniques can communicate the most effective data stories.

McCandless, D. (2014). *Knowledge is beautiful*. William Collins.

This book by data journalist David McCandless demonstrates in a clear and engaging way how good visualisations can reveal compelling and powerful statistical narratives.

7

THE STORY WAITING TO BE TOLD

Chapter Overview

Introduction 200

The opposite sex? 201

The gender pay gap 202

Data still matters 203

Measurement still matters 205

'Let's talk about sex baby' 205

The importance of context when exploring data 207

Beyond the double standard: telling stories of
sexual difference 209

Men estimate – women count: a gender difference to
recollecting 210

Trigger warning! Statistics in the media 210

The transient nature of the news media 213

Butter is good/bad/good/bad/WTF for you! 213

The fine line between mistakes and misleading 216

The devil's in the detail! 217

Purveyors of fake news! 217

Further Reading 219

Introduction

What do we mean by narrative? Well, it is relatively simple. What we want to suggest, and have done so throughout this book, is that behind the numbers, data and statistics there is a story waiting to be told. This is narrative. When using data and statistics, it is not always obvious or clear which story can be told as there are so many possible directions that a narrative can take. It is therefore the job of the social scientist or social science student to develop this narrative and, essentially, bring the data to life. You might wonder why we are now talking about telling stories and developing a narrative: *that cannot be right, as this is journalism*, we hear you say. Well, journalists also tell stories although they tell them in a different way from the social scientist or social science student. But you would be correct to say that there is some crossover in what we do. We both use evidence, speculate on likely causes or outcomes, while also considering the audience we are trying to persuade. Of course, there are also differences too. We'll explore this later in the chapter. Suffice to say, the narrative is a way of trying to explain something in the social world, in the same way as social theories are also trying to explain something. In this instance, the narrative is developed from the descriptive statistics, data analyses and social theory combined.

Throughout this book, we have told many different stories. Can you recall them? Two personal favourites of ours are sex/gender (as in our biological or socially constructed version of our sex roles) and sex (as in the act of having or not having sex). Often, we are looking at sex/gender differences between men and women. Remember in Chapter 3 on measurement, we suggested the absurdity of the notion of 'opposite sex' between men and women that focus on the few differences between us but ignore the many similarities that exist. These are all stories that evolve from social and/or cultural notions of 'sex' and 'gender'. Or when we use the concept of sex, we might be exploring attitudes towards 'having or not having sex' here looking at the number of times people engage in sex acts. There are a lot of normative behaviour codes surrounding sex, making it a widely researched area. Often, having sex is seen as the norm while also quite heavily stigmatised, particularly if you are deemed to have 'too much', especially if you are a woman in most cultures; while not having sex has a discourse of abnormality that surrounded the labels 'asexual' or 'celibate'. The word 'celibate' has an interesting narrative in that it is a 17th-century Latin word for 'unmarried'. Clearly, the expectation here was that unmarried people should not engage in sex, whereas those who were married could. Now however, it refers to someone who abstains from sex. Through collecting evidence, 'counting' for example, the number of people who report being celibate allows researchers and writers to develop a narrative around 'not' having sex. Indeed, there are some emerging trends which suggest that the under-25s in many nations are having less sex than previous generations, and in some nations, like Japan, celibacy is so common amongst the under 30s that it is deemed a 'social problem' (Ghaznavi et al., 2019). However, as suggested in Chapter 3, there must be consideration of how we could measure such a concept. Should those under 16 years old be included? What about people who want to have sexual relationships but are unable to because of medical problems they have? Do we include all sexual acts or only some, as 'proper sex'? The Natsal-3 survey that we used in Chapter 2 is a good example of a survey attempting to measure sexual activity; go and have a look at their website (http://www.natsal.ac.uk) to find out more about how they approach the subject.

This is the beauty of social science, we don't just count; we try to measure and then explain, remembering that the issues we engage with are never as simple as they at first appear. It seems, however, that 'number' does not always sit centre stage when theories are developed. This can lead to too much attention on theorising for its own sake, making theories too abstract (and often difficult to interpret) and not grounded in evidence – that is, robust data. Let's start unpacking this with example of gender.

The opposite sex?

Gender, although increasingly complex of late, is often considered to be made up of a relational construction between 'masculinity' and 'femininity'. 'Sex' (biological sex that results from the XX or XY chromosomes), is different in that in most cases, sex is set at birth. 'Gender' on the other hand is argued to be socially constructed (Oakley, 1972). An easy way to evidence this is to look at how gendered normative practices differ within and between cultures. They also change over time. Just think for a moment how flamboyant the men dressed in the Elizabethan era. Such dress is how they defined their masculine identity in Elizabethan times; today the triangular beard and large codpiece has been replaced with hipster beard and skinny jeans.

An important narrative regarding both 'gender' and 'sex' is that of inequalities, whereby men and women are treated differently based upon their assumed sex. It is strange to think that with so many similarities between men and women, that the phrase 'opposite sex' has such cultural importance. That is not to say there are no differences, as there clearly are. It just seems that the differences have been exaggerated out of proportion and the similarities minimised. This has not always been the case and our assumed sex differences were not considered humanity's defining feature.

Did you know, for example, that in the medieval period, there was only thought to be one sex, the prototype we now consider as 'male' (Hawkes, 1996). The story goes that being female did not rest on the position of a vagina (or penis if male) but rather, half of the sex organs were just the same with half of mankind having external sex organs and the other half having internal sex organs. Interestingly, lack of external female sex organs was actually considered as a distinction of incompleteness (Hawkes, 1996). Unlike the biological variations between the sexes, the vast majority of differences relate to differential treatment of men and women; and these are very much within our control (Oakley, 1972). Take pay as an example.

The gender pay gap

In Chapter 5, we explored the gender pay gap. Did you know that since the 1970s, it is actually illegal to treat men and women differently in the UK (and in many other countries)? This is especially so around pay. Yet, as of writing this in 2019, the gender pay gap still exists (Olsen et al., 2018). Throughout history, it seems that women have been paid less than men, even when they are doing the same or essentially the same job as a man. Reasons given to justify this were that women were weaker and intellectually inferior to men, resulting in them not deserving equal pay. While we have made great strides of late, such views persist in some, for instance only back in 2017, Polish MEP Janusz Korwin-Mikke suggested this was the case. While we may dismiss the rants of an out of touch white heterosexual man, we still need to explore why the gender pay gap exists. Just reviewing the data for such differences is not enough and there needs to be a convincing explanation for this (remember, a theory is just a way to explain something). A quick answer to this would be 'male privilege' but this does not really explain in sufficient depth. Instead, we can also try to explain the processes that allow this. For instance, Williams (1992) introduced the notion of the **Glass escalator** – the existence of an unofficial fast track for white heterosexual men entering what is often deemed or was traditionally a female-dominated profession, such as teaching, nursing and social work (Connell, 2013). This is just one of many ways we could use to explain why men and women, at a society level, earn different amounts of pay. We could also look at why women earn less than men. Loden, in 1978, coined the phrase **Glass ceiling** alluding to an invisible glass ceiling that prevents women from climbing their career ladder (Daniels, 2013). Researching the numbers of women who hold the most senior posts at key organisations, such as the BBC or many top businesses, does reveal that those at the top are mostly male (e.g. see Hospido et al., 2019). However, what the notion of the glass ceiling does not take into account are those women who for one reason or another do not have a career path and are just doing all they can to make ends meet. For these women, we could turn to Berheide (1992) who offers us the explanation of the **Sticky floor**. These are types of jobs that are low-paid, low-skilled and low-prestige jobs typically held by women that have little opportunity for career progression and development (Berheide, 2013). Added to the sticky floor metaphor is the '**5 Cs of women's work**', where women are said to 'cluster' around catering, cleaning, clerical, caring and cashiering.

Data still matters

What these ideas have in common is how they try to explain why, on the one hand, men receive higher pay than women; and conversely, why women are paid less than men. What is interesting about these ideas is how they are all formulated on robust data that reveal how such differences exist. Yet, in some respect, it is not made immediately obvious that this is the case as the theory takes centre stage, meaning we do not always need to focus on how this is linked to the data. But think about it, we need to link back to the data to show the average wage of men and women (in comparative jobs and sectors) to make such comparisons and then to develop the theory. If you go to the UK's (or any other national statistics office) ONS website (https://www.ons.gov.uk/) and look for the gender pay gap, you will be confronted with an array of quite complicated statistics that clearly show how men are consistently paid more than women. With a little time and effort, it is possible that most of us would be able to glean some information about this. However, if you think back to Chapter 3 where we suggest how it is helpful to break down how social researchers use variables into (1) classification, (2) Counting, (3) measuring and (4) explaining, it is the explaining aspect of the process where data and theory should be linked rather than unduly focusing on one or the other.

Yet when we are carrying out our analyses and generating descriptive statistics, the findings in and of themselves are not the narrative and viewed in isolation do not really tell us very much at all. This is why it is up to us to link them to social science theory to better explain them and their trends or patterns. In essence, the numbers are a story waiting to be told. What the authors of the above theories on the gender pay gap have done is to put the 'explain' aspect centre stage to the point that the counting (data) aspect can even seem unnecessary. The way they have explained this or turned it into a **narrative** resonates so much that we almost do not need the data. The narrative is therefore possibly the most important part of the process in that the researcher is attempting to explain why, in the case of gender pay gap, such inequalities exist.

Where and when the story stops is very much up to you. To be thorough, as is our job as social scientists, we should always try to tell the most complete story possible. That might be we look beyond our personal and/or social context and explore what gender inequalities might look like elsewhere. For example, international comparisons, as illustrated in Sadoff (2019) showing the ranking of global gender inequalities.

What Figure 7.1 shows is a magnitude of gender-based disparities between 144 different countries. It shows the top 10 with Iceland having the highest gender equalities score. The UK is ranked at Number 15 with Yemen being ranked the lowest.

Figure 7.1   Gender-based disparities found across the world

Source. Global Gender Gap Report 2018, World Economic Forum – http://reports.weforum.org/global-gender-gap-report-2018/shareable-infographics/

Box 7.1

Pause for Thought

What story could you tell just using the data of gender equalities in Figure 7.1?

Measurement still matters

Figure 7.1 tells a very incomplete story, and the curious amongst you will already want to know how this was being measured. For that, you could go to 'The Global Gender Gap Report 2018' (http://reports.weforum.org/global-gender-gap-report-2018/shareable-infographics/) that has all the necessary details needed to show how this was measured and explained. What is important here is that through measuring what is considered to be gender inequalities, such as receiving different levels of pay, or being prevented from doing certain activities, such as owning property or voting, the authors have been able to use both data and sex/gender theory to tell this incredibly important story. A story that looks at the way that women are, on the one hand, being treated equally and fairly in some countries; but on the other hand, how across the world we have some way to go to improve the treatment of over half of the global population.

'Sex' and 'gender' might at first glance appear to be an easy narrative to develop when using data as we are often looking at differences. Yet, as discussed in Chapter 3, account should be taken of the complexity of any idea we are working with. We now know, for example, that there are ongoing debates about the number of sexes with some suggesting that it should no longer be limited to two, formulated upon various biological markers (Fausto-Sterling, 2000; Mak, 2012). The idea of an intersex person, that is someone who does not fit in neatly with the idea of there only being human beings made up of the XX or XY chromosomes is starting to gain traction (Sanz, 2017). Similarly, with gender, as this is a social construction anyway, we can conceivably have any number of genders or, indeed, none at all. What is perhaps important here is that we work with the data we have and develop any stories we can from that data. That is not to say that the narrative cannot explore any of these complexities as this critique lay at the heart of social science and is the motivator for its continued progression. However, if your data, for example, is limited to male and female, then conceivably, we are only able to report on sex and/or gender.

'Let's talk about sex baby'

Now that we have discussed the notion of 'sex' that being male/female or other, we can now move on to the other type of sex, as in the act of having sex. Again, this is another interesting social science subject that at its heart has data that underpins the stories, narratives and theories. Think for a moment, the importance of numbers linked to the idea of sex. Just as with gender, the act of 'having sex' has a biological and socially constructed component. One recurrent theme in research around sex is the numbers of sexual partners someone has over a lifetime; and how this differs for men and for women. We are lucky in that over the last three decades, research has been carried out into this subject area by The National Survey of Sexual Attitudes and Lifestyles (Natsal) – remember we looked at Natsal-3 in Chapter 2? The first findings were reported in 1990–1991 (Natsal-1); the second survey findings were published in 1999–2000 (Natsal-2); with the latest results published in 2010–2012 (Natsal-3). Figure 7.2 shows the number of lifetime partners reported by men and women, aged 16 to 44 years old, in all three Natsal surveys.

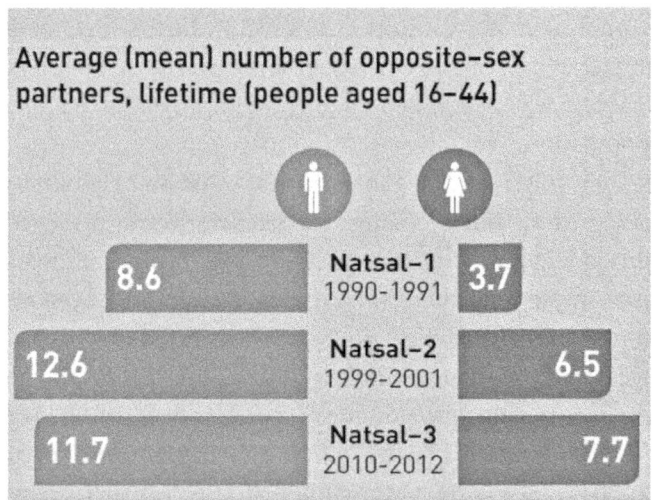

Figure 7.2 Average number of lifetime sexual partners of men and women in the UK aged between 16–44

Source. http://www.natsal.ac.uk/media/2102/natsal-infographic.pdf.

Box 7.2

Pause for Thought

Figure 7.2 shows a bivariate analysis using averages in three different time periods. Looking at the data, what story are you able to tell?

The first thing of note is how there is an obvious gendered element to the number of times people report having sex; with men reportedly having more partners than women. For instance, men state they have had 11.7 opposite sex partners compared to women who have had 7.7 as of the last count, using Natsal-3. The reporting of such differences between men and women is clearly consistent across the three decades. The question is why this might be – what's the story? Thinking about possible explanations that could account for differences in the number of sexual partners in a lifetime, we could turn to biological and neurological differences (Carvalho et al., 2013). The story goes that men want sex whereas women want commitment (McCabe, 2005). They do sound plausible looking for differences in hormone levels and/or brain activity. What is interesting about this narrative is how it asserts that we know more about the human brain and biology more than we actually do (Mitricheva et al., 2019). Mitricheva et al. (2019), reviewed 61 neuroimaging articles, consisting of 3723 adult participants, and they could not establish any differences between men and women of all ethnicities and sexual orientations. How then do such differences enter into the lexicon of our understanding of gender? That is not to say that such explanations are without merit. The issue here is, as social scientists it is important that other explanations are also considered. It is also important that we recognise when theory has overreached itself and is, in essence, making connections without the necessary evidence.

The importance of context when exploring data

When we try to *explain* using data, we need to include other ideas to ensure we tell as complete a story as possible. These include some of the many theories that have been developed to explain social phenomena, such as differences in attitudes towards sex. These will be explored in more detail below, but it is worth remembering that such theories are not developed in a vacuum, which means we should also take the social context into account when exploring them. For instance, sexual attitudes in the 1950s, in the UK and the USA, were very different from what they are now. Premarital sex was shunned as was any 'other' expression of sexual identity. This clearly would impact on the type of questions that might have been asked (if at all) about the sex lives of men and women. They would have also influenced the types of answers given with *subjects* possibly providing socially desirable answers with 'good' behaviour being over-reported compared with behaviours considered 'bad' (Frey, 2018). That means when considering how data will add to the narrative, socially desirable behaviours will be determined by the climate of the time when it was collected. As such, when developing the narrative that will accompany data, these are just some of the issues to be considered. Look at the cycle in Figure 7.3.

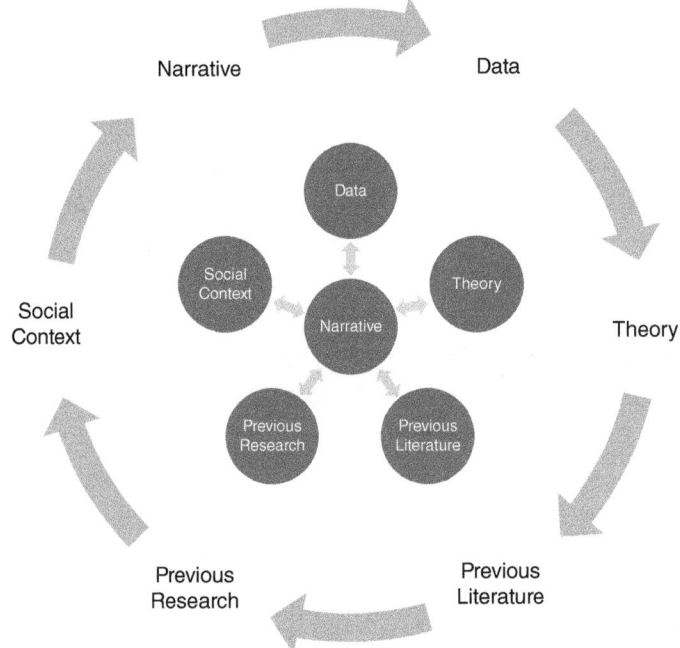

Figure 7.3 The building a narrative cycle

In some respect, the narrative is at the heart of the research endeavour while at the same time an important part of the process; hence, the use of two diagrams as neither one seems sufficient enough to reveal the complexities of the process. We have used six key areas to consider:

- Previous research
- Data
- Previous literature
- Social context
- Theory
- Narrative

Each of these add to the story waiting to be told. The distinction between the different elements are somewhat artificial and they have much crossover between them. As quantitative research is deductive (top-down), it starts with a good grounding in existing literature and prior research. It is with such evidence that we begin to build the story we want to tell once we start to use the data. Using the previous example of the reported differences in the number of times men and women engage in heterosexual sexual relations, let's now explore how all these different elements can go to make up the narrative. Let's start with previous research, previous literature and theory.

Beyond the double standard: telling stories of sexual difference

There are many books and journal articles published that explore and explain the gender double standard in attitudes and actions toward sex (e.g. see Hawkes, 1996; Kettrey, 2016; van Hooff, 2015; Zaikman & Marks, 2014). The double standard is when men and women have to follow different cultural normative practices that reward men for their promiscuity, while at the same time stigmatising women who engage in the same sexual practices. Depending on the discipline, they will be offering biological or social explanations for such differences (e.g. see Lippa, 2009, who carried out research across 53 different nations). Both forms of explanation could be implicated; and the acceptance of one does not mean the rejection of the other. As it turns out, Lippa did not find very much evidence of gender difference in the meta-analysis carried out. Yet we can see the data consistently shows differences between men and women in the reporting of frequency of engaging in sex acts, social scientists need to try and explain why this might be.

One issue to consider is that the type of data used when reporting such differences is observational (rather than experimental), meaning it will always be open to scrutiny. This is why the social context should also be taken into account. We know, for example, that while sex still carries significant cultural baggage, especially around notions of monogamy, the stigma has reduced significantly since the 1960s and the onset of the contraceptive pill. This might account for the apparent increase in the number of sexual partners over the life course shown in figure 7.2 It would not, however, explain why men and women continue to report differences in the number of partners.

Rather than biological explanations, we could also turn to how gender is socially constructed. Remember, this is not about biological sex, where such differences are biologically determined, but rather, gender that is feminine and masculine normative practices that are specific to a particular culture (in this instance, the UK). Such differences as reported in the Natsal data are illogical as they should be equal (as the old saying goes, 'it takes two to tango . . .'). Two possible explanations for such difference were explored statistically by Einon (1994). Here, she wanted to see if hyper-sexualised women and prostitution were skewing this and other data reporting such differences between men and women. Using statistical models, she was able to discount both explanations. Put simply, such outliers did not exist for either men or women that would be needed to account for such differences. She reported that such differences are therefore inaccurate.

So, ruling out that men and women do actually have different numbers of sexual partners over a lifetime, what else can we use to explain this other than the theory of gender? Again, turning to the notion of the double standard, could this be put forward to explain *why* men and women report different numbers of sexual partners reported in all three Natsal surveys? Could it be that women under-report due to the social stigma of being a *too* sexually active woman? Could it be that men over-report the number of sexual partners to bolster their masculine identity (Mitchell, 2018)? All seems very plausible. You might note that thus far, we have attempted to explain such differences stemming from very little data showing that year on year, men and women report different numbers of sexual partners over a lifetime. To help us better understand why this might be, we have developed a number of stories acknowledging that all are deficient. We could go further and look beyond sexual practices in our attempt to explain this phenomenon.

Men estimate – women count: a gender difference to recollecting

Again, we are looking for possible gender differences that could explore such differential reporting of sexual partners. Another possible explanation is provided by Mitchell et al. (2018). Using the Natsal-3 data, they were able to show that the differences came about because of differences in how the numbers were calculated. They concluded that men estimated, often higher, the number of sexual partners, whereas women more precisely counted the numbers and indeed kept a tally. While they could not fully eliminate the differences, they did reduce them considerably. What is interesting is that the differences in sexual partners between men and women could still be an artefact of gender but not necessarily in the way it first appears. It might be a simple matter of women keeping a better score of the number of sexual partners they have had. It might be that a sexual encounter is more meaningful for women, whereas men quickly move on. Of course, this is all speculative but in a way that is trying to provide a better explanation to the observed differences in the data. As this issue is explored in more detail, more and more possible explanations will add to this story. This is an active rather than a passive process that the social scientist performs that will bring the data to life. The important thing is that this exploration takes place rather than taking the data at face value. As suggested previously, the story is waiting to be told.

Trigger warning! statistics in the media

We hope the above section will encourage you to develop your own narrative around the social issues you are currently exploring. It is all about the story rather than just the data. You might however ask how this differs from that of journalism and how news media tell their stories. This is a good question and one that is fraught with difficulties when trying to answer. It is true that both the news media and social scientists use evidence (data) to develop their stories, but there are key differences. Let's try and explore some of them. The first thing of note is those working in the news media might not be properly trained to report fully on stories leading them to overgeneralise and take them out of context. Take the headline below:

> Average child will receive £56 worth of Easter eggs and consume 8,000 calories of chocolate this weekend. *Daily Mail Online* (Davies, 2015)[1]

Box 7.3

2-Minute Recap!

Set a timer on your phone for 2 minutes; in that time, can you decipher what is wrong with this headline? Once you have done this, can you rewrite the headline as it should be?

Did you see how the above headline has assumed the child to be the average measure (which it quite simply cannot be) rather than the average consumption of Easter eggs by children? If you did, well done! Of course, the headline should read as follows:

> Children will receive £56 worth of Easter eggs and consume an average of 8,000 calories of chocolate over the Easter weekend.

There are many instances when the news media report statistics and numbers correctly, for instance, when reporting soccer matches. They will tell the final score, the stats of player positioning, shots on target and so on. They seem to get this type of statistic correct but then on issues such as health, diet and so on, they can so monumentally get it wrong.

Take a look at some of the examples in Chapters 4 and 5 where we explored the reporting of percentages, frequencies and averages. You hopefully know that when reporting averages, it is important to report the correct average. Or when reporting percentages, it is important to also give the frequencies. Sadly, many media outlets do not adopt this approach. Take the headline below:

> 90% of teens unhappy with body shape. *Daily Mail Online* (13 August 2004)

---

[1]The *Mail Online* has an extensive back catalogue of their articles, making them an easy target. We used *The Mail Online* because of this accessible back catalogue; all tabloid newspapers are poor at reporting statistics and even the so-called quality broadsheets can have lapses in reporting quality.

Box 7.4

Pause for Thought

Before reading on, can you identify what other information is needed in the above headline?

The news story goes on to point out how:

> Nine out of 10 British teenage girls are unhappy with their body, with mothers appearing to be responsible for passing on their own insecurities, a new survey shows. Only 8% of the 2,000 girls questioned for the poll said they were 'happy' with their appearance, while 87% said they were 'unhappy'.

With the example above, we have provided a little more of the story to show that upon further reading, the headline is gender specific but does not state this. In this instance, it is poor reporting of statistics rather than getting them wrong. However, look at these other examples below of poorly reported statistics:

> The average parent estimates they make 13 suggestions to their child every week during the summer break to help keep their child or children entertained and the average child starts to get bored just after 1.30 pm on a typical day during the summer break. *DailyMail Online* (19 June 2017)

Did you identify what was incorrect with the reporting of these headlines? Of course, you did. What is meant by 'average parent' and 'average child'. Sometimes, the way the statistics are written seems fine but upon closer inspection is not. Let's take a look at a story on the gender pay gap between male and female hospital consultants taken from *The Independent*:

> Meanwhile, full-time female consultants earn an average annual salary that's nearly £14,000 less than what an average male consultant earns, the investigation found. That represents a pay gap of 12 per cent. *The Independent* (16 February 2018)

Can you spot what is wrong with it? Again, it is incorrect reporting of the average. It starts correctly as it discusses average annual salary but then suggests an average male consultant. What is an average male consultant? He is clearly male but is he 50 years old or 60? The mistake is to suggest there is an average male consultant when they are really trying to make the point that the average salary of men and women is different (it is salary that is the average, not the person).

OK, last one.

> The average Brit drinks 876 cups of tea a year – enough to fill two bath tubs. *Metro* (Goorwich, 2015)

Apparently, the 'standard' length of a bathtub is 5 feet 6 inches or should that be 1.7 metres. Of course, the headline does not deal with the differing size of the bathtubs or the non-standardisation of cups used from which to drink the tea. So much detail is missing that it hardly seems worth engaging with the headline. Although, this is about the national identity of British people so is an important issue. Perhaps the headline of *Brits still drink loads of Tea, especially older people* is quite dull but much more truthful. Or perhaps the headline stating that *British people drink on average 2.4 cups of tea: a day!* is so much less sexy. Yet in essence, the media are performing the same role as is suggested by this chapter – telling a story. A key difference is we, as social scientists, are more often than not trying to add to a story in the pursuit of generating new knowledge and understanding. A journalist and media organisations work at a much faster pace and quickly move on to the next story.

The transient nature of the news media

Another key difference of the news media is the need to sell advertising space or 'clicks' on stories. This is a highly competitive market in which they operate, resulting in the need to quickly grab the attention of the reader with some catchy headline that they hope will spark enough interest in their story for the reader to buy their newspaper or be in some other way exposed to their advertising. As well as them sometimes getting their stats wrong (which we all do from time to time), another issue relates to their oft transient views on the news stories they report. It seems they can change views from one day to the next. Of course, they are reporting the news, and if that changes then so does their headlines. Take health and diet stories for example. They seem to pick up on a news story, run with it, state the significance of it to our everyday lives, and before the general public even have time to digest it, it is dropped for the next big headline. This can leave many in a real state of confusion. Take the reported benefits and risk to health of that most simplest of food items: butter!

Butter is good/bad/good/bad/WTF for you!

Below are some headlines from the *Daily Mail* between 2009 and 2019. The stories go from eating butter is bad for you to eating butter is good for you. There are lots of comparisons to other things that might be good or bad for our health, such as lard, sugar or avocados ending with the most recent telling us to STOP worrying!

Google search 'mail online butter bad for you' conducted on 29 April 2019:

- 19 January 2019

Eat what you want and STOP worrying! New book claims sugar ISN'T toxic, low-carb diets AREN'T healthier and avoiding dairy is POINTLESS

https://www.dailymail.co.uk/health/article-6610539/A-new-book-everyones-talking-urges-eat-want-STOP-worrying.html

- 28 December 2018

Cardiologist slams 'incorrect' advice from the World Health Organization urging people to replace butter and lard with healthier vegetable oils in 2019

https://www.dailymail.co.uk/health/article-6535725/Cardiologist-slams-incorrect-diet-advice-World-Health-Organization.html

- 22 October 2018

Leading medical journal is accused of having a 'pro-BUTTER bias' as more than 160 academics demand its editor stops publishing 'odd' studies

https://www.dailymail.co.uk/health/article-6302639/Leading-medical-journal-accused-having-pro-BUTTER-bias.html

- 26 February 2018

Revealed: How lard is actually GOOD for you (and it can even be healthier than butter)

https://www.dailymail.co.uk/femail/food/article-5427303/Food-Unwrapped-reveals-lard-healthy-butter.html

- 26 April 2017

Butter and cream will do you no harm (if you walk for 22 minutes a day): Doctors say avoiding saturated fats in these products does nothing to reduce heart disease

https://www.dailymail.co.uk/health/article-4445718/Butter-cream-not-harmful-walk-22-minutes-day.html

- 15 June 2017

Why you should switch from butter to margarine: Simple change could be as good as statins for your heart

https://www.dailymail.co.uk/health/article-4607854/Why-switch-butter-margarine.html

- 16 June 2017

Don't use coconut oil, it's 'as unhealthy as beef fat and butter', American Heart Association warns

https://www.dailymail.co.uk/health/article-4611736/Coconut-oil-unhealthy-beef-fat-butter.html

- 20 August 2016

Trendy avocado spreads are 'less healthy than butter': Unhealthier oils outweigh the beneficial ones, says leading nutritionist

https://www.dailymail.co.uk/news/article-3750007/Trendy-avocado-spreads-healthy-butter-Unhealthier-oils-outweigh-beneficial-ones-says-leading-nutritionist.html

- 29 June 2016

Butter is NOT bad and doesn't raise the risk of heart disease, major study claims

https://www.dailymail.co.uk/health/article-3666211/Butter-NOT-bad-doesn-t-raise-risk-heart-disease-major-study-claims.html

- 10 February 2015

Butter ISN'T bad for you after all: Major study says 80s advice on dairy fats was flawed

https://www.dailymail.co.uk/health/article-2946617/Butter-ISN-T-bad-Major-study-says-80s-advice-dairy-fats-flawed.html

- 7 February 2013

At last, the truth: Butter is GOOD for you – and margarine is chemical gunk

https://www.dailymail.co.uk/health/article-2274747/At-truth-Butter-GOOD--margarine-chemical-gunk.html

- 6 February 2013

Swapping the butter for margarine 'may be bad for your health'. U.S. scientists claim polyunsaturated fat 'doubles heart risk'

https://www.dailymail.co.uk/health/article-2274168/Swapping-butter-margarine-bad-health.html

- 21 March 2009

The great butter mystery: It's natural, so why is it no good for you?

https://www.dailymail.co.uk/health/article-1163680/The-great-butter-mystery-Its-natural-good-you.html

Phew! Talk about confusing. I think I'll take my chances on eating butter. The question is, 'Are the stories the news media report incorrect?' The answer of course is not that simple. They may or may not be. They are often not out and out lies but rather taken out of context and not produced with the necessary caveat that would accompany them if they were reported in academic books and journals. It is not just the fault of the news media but also the type of nutritional research they report. Often, Ioannidis (2018) reports, this is observational research whereby participations are asked about their dietary habits. As such, these are not random controlled trials and as such will and should have this element of doubt; if you are interested in the differences between observational and experimental approaches, then read Volume 4 (*Experimental Design*) in this series. Similarly, whereas academics *should* have time to consider possible alternative explanations, news media do not have this time before they have dropped the story, sometimes in a matter of days, and moved on to their next.

Box 7.5

Reflective Exercise

Thinking about other health issues which have been reported as good/bad depending on the era, by the new media, reflect on the following list:

Different types of diets have been reported as the 'next big thing'.

Red meat is good/bad for you.

Skipping breakfast is good/bad for weight loss.

Think about how the media have approached each of these health topics over the years; you could always research them online and see how many times search issues have been 'healthy/not healthy'.
    Can you think of any other health advice that the media has reported in a similar way?

The fine line between mistakes and misleading

Mark Twain, the American writer and humourist, is reported as saying 'Never let the truth get in the way of a good story'. As well as times when the news media simply get things wrong, there are also times when they appear to set out to deliberately mislead.

Fortunately, here in the UK we have the Independent Press Standards Organisation (IPSO) to oversee when this happens. An interesting example of this recently occurred involving *The Sun* newspaper; in 2015, reporting on a survey they had commissioned stated on their front page that nearly 1 in 5 Muslims in the UK had sympathy *for* those who went to fight for ISIS in Syria. This was presented as a shocking fact and that they had the necessary data to back up such a claim. However, when the headline was scrutinised, it soon became clear that the newspaper had played fast and loose with the facts and as Mark Twain might say, had not let the truth stand in the way of their good story. Let's examine this in a little more detail.

The devil's in the detail!

IPSO found that the headline was quite misleading. The first issue identified was how the question in the survey did not explicitly state 'fight for ISIS' but actually said 'who leave the UK to join fighters in Syria' meaning some of the respondents could have gone to Syria to fight *against* ISIS. IPSO stated that *The Sun* had distorted the poll with this reporting. The second big issue was how the Sun suggested that respondents had sympathy *'for'* those leaving the UK when the actual question asked sympathy *'with'* those leaving the UK which has quite different connotations. The final issue related to how only reports of the views of Muslims were reported when there was evidence of non-Muslim people holding similar views, in similar numbers. This illustrates that not only can the media 'get things wrong' when reporting data but also can 'mislead' using data (IPSO, 2015). This leads us to recount the oft quoted phrase of 'lies, damn lies and statistics'. We reject this idea, as numbers and statistics don't lie. People lie and can use numbers and statistics as their weapon of choice to justify the argument being made.

Purveyors of fake news!

What social scientists and novice researchers should consider when exploring almost any issue is how much of what we work with is observational data. If we could carry out large-scale experiments on the dangers of smoking while pregnant, benefits of eating breakfast, or the impact of butter on the health of the individual, we might (only might) reach a much firmer conclusion. But as this is not possible, we are then left with the need to introduce caution and uncertainty into the stories we produce. Statistics are used by the news media to add credibility to their story, and it is our job to check and call them out when they make false claims. In this news- and data-saturated world in which we live, why do we easily believe the stories we are being fed without checking them out? Part of the reason this happens with such ease is the lack of statistical literacy we have in the UK and across the Western World (Scott Jones & Goldring, 2017). In schools, colleges and universities, there seems to be much more focus on our writing skills, which clearly are important but then so are data and numeracy skills. Yet as we live in an increasingly data-rich society, our need to more fully scrutinise the evidence we read in the traditional media, such as newspapers, and also the new 'social' media, such as Facebook, Instagram and Twitter, has never been more important.

We have the skills (social science skills of research design), technology (access to the web and academic journals) and love of all things internet (going by the amount of time being spent using it, which for the average teenager is around 7200 seconds per day). Yet, for whatever the reason, many of us like, share and retweet with little or any thought for accuracy. In essence, *we* are the purveyors of **fake news**. This goes against the science doctrine of falsification, whereby we start with the premise that the claim is false. As it stands, we seem to regurgitate claims we already agree with. It might at first glance seem that this is a new phenomenon brought about by the new social media, such as Facebook, Twitter and Instagram. This is incorrect as this has a much older heritage seen in how particular 'types' of people would read specific newspapers in a tautology, whereby the views of the readership are represented by the newspaper which is then read by the readership with that particular worldview. So, before we take an accusatory look at the rest of the world, let's look internally and develop our own statistical literacy and start to challenge when we see or hear nonsense stats (but let's not contribute to it).

What can be done? As well as being statistically literate, there are online tools that can help us check facts. For instance, the Social Research Association lists a host of fact checking sites – http://the-sra.org.uk/sra_resources/fact-checking/.

For instance, one such site, 'Full Fact' suggests it is an independent fact checking charity in the UK. We are in the strange situation of being exposed to so much fake news while at the same time having the necessary tools to check the stories we are reading in the media (old and new). Perhaps, it is our civic responsibility to check before we share.

In this chapter, we have explored the importance of developing a story when using descriptive statistics. In essence, the data and findings from surveys is not enough to persuade, making it 'our' responsibility to tell the story waiting to be told. The stories we tell, however, must always contain the caveat that we are only offering one possible explanation and other explanations are available and viable. We should not shy away from the complexity of what we do as social science researchers or students. Stand proud, you are studying a highly complex subject made harder by the vast array of possible stories we can tell. The world in which we live is far from black and white. There are so many shades in between. This is our core business in exploring that, which inhabits the centre ground, the grey areas of life. By doing this, we take small steps in our understanding of the social world. By using what is easily understood to help us with this endeavour, descriptive statistics can help us do this better.

Chapter Summary

- The narrative is developed from the descriptive statistics, data analyses and social theory combined.
- This is the beauty of social science, we don't just count; we try to measure and then explain, remembering that the issues we engage with are never as simple as they at first appear.
- Findings in and of themselves are not the narrative and viewed in isolation do not really tell us very much at all. This is why it is up to us to link them to social science theory to better explain them and their trends or patterns.
- When we try to *explain* using data, we need to include other ideas to ensure we tell as complete a story as possible. Six key areas to consider are as follows:
  1. Previous research
  2. Data
  3. Previous literature
  4. Social context
  5. Theory
  6. Narrative
- What social scientists and novice researchers should consider when exploring almost any issue is how much of what we work with is observational data. Thus, social context should also be taken into account. Do not take the data at face value.
- Statistics are used by the news media to add credibility to their story, and it is our job to check and call them out when they make false claims.

Further Reading

Best, J. (2013). *Stat spotting: A field guide to identifying dubious data.* University of California Press.

This is an interesting book that will get you thinking about how the media uses (and misuses) quantitative data. It is useful, as this book is not a critique of statistics but rather debunks how they are often misreported by media outlets.

Blastland, M., & Dilnot, A. W. (2007). *The tiger that isn't: Seeing through a world of numbers.* Profile Books.

This book is a fantastic and engaging review of how the media, government and big business use and misuse statistics to create their own **statistical narratives**. It is packed with great examples and will certainly improve your statistical literacy by the end of the book.

Mosher, C. J., Miethe, T. D., & Hart, T. C. (2011). *The mismeasure of crime*. Sage.

This book is really helpful in showing the complexities involved with measuring anything in the social world, but specifically, the measurement or mismeasurement of crime. The media often sensationalise crime, but in this book, the authors cut through all the lurid headlines and misuse of statistics to reveal how data can be presented in a way that does not enlighten but rather misleads the reader.

Lidskog, R., Berg, M., Gustafsson, K., & Löfmarck, E. (2020). Cold science meets hot weather: Environmental threats, emotional messages and scientific storytelling. *Media and Communication, 8*(1), 118–128. http://doi.org/10.17645/mac.v8i1.2432

This journal article is a useful exploration of how the art of storytelling in the sciences can be transformative and better influence those in positions of power, policymakers and decision takers. It is no longer enough to present the 'cold' distant facts; for science to be an agent of change, it needs to produce a compelling and persuasive narrative, in this case about climate change. The trick, they argue, is to strike the right balance between emotional appeal and scientific authority.

GLOSSARY

5 Cs of women's work: The 5 Cs of women's work refers to the low paid low skilled jobs of catering, cleaning, clerical, caring and cashiering that gender structures often limit women to doing.

Administrative data: It is data typically collected by organisations, such as governments, private companies, charities and so on for the purpose of record keeping about interactions with their user population. These can include, for example, school records, membership records, electoral registers as well as registrations of births, deaths and marriages. While not collected for research purposes, such data is frequently used in research.

Bivariate analysis: Analysis using two variables. Bivariate analysis can be used to compare data such as gender (male/female) and average annual earnings or to look at relationships between two variables, such as the link between car usage and air pollution.

Categorical variable: A categorical variable is one that measures non-numeric groups (categories) such as male/female, school qualification or ethnicity. Categorical variables can be either nominal or ordinal. Gender and ethnicity are categorical nominal variables, whereas qualification type can be considered as either categorical nominal or categorical ordinal. This is because it can be ordered in a logical way from the lowest to the highest qualification (hence categorical ordinal).

Data cleaning: This is data that has been 'cleaned' prior to analysis.

Counted: This is a term used to determine the number of items or frequency that something occurs.

Cross-tabulation, crosstab or contingency tables: These are tables that allow researchers to present descriptive statistics within a bivariate analysis. The tables facilitate the initial exploration and summary of associations between two variables. Using two variables allows for patterns and relationships to be described such as difference in time spent on social media between males and females.

Data cleaning: It is the process whereby data is prepared for analysis. This typically involves checking and removing errors, plus further manipulations of the data to ensure that it is suitable for the planned analysis.

Data familiarisation: It is an important aspect of exploratory data analysis where the researchers use a range of data visualisation techniques to explore their data prior to any formal analysis.

Data management: It is a complex process that involves all aspects of managing data for a research project. It includes data accessing, ethical controls, data cleaning, data storage, versioning and archiving of data.

Descriptive statistics: Descriptive statistics describe and summarise data taken from a sample (the 'n') of a wider population (the 'N'). Descriptive statistics can be both univariate and bivariate. No generalisations can be made from descriptive statistics back to the wider population.

Exploratory data analysis (EDA): Developed by Tukey, EDA is a technique to explore data which looks for insights, patterns and trends without formal inferential or significance testing.

Fake news: It is an untrue story that claims to be factual. They are often sensationalised and widely shared on social media.

Frequency tables: A frequency table shows the distribution of the possible answers occurring from categorical data. They include just one variable and are used to summarise the data. This could be the distribution of males or females, different ethnic groupings or the type of qualifications of those included in the data set. In IBM SPSS, the frequency table will also include the percent, valid percent and cumulative percent.

Frequentist approach: This is the dominant approach to statistical analysis and involves null hypothesis significance testing of suitably sampled and cleaned data. This approach is increasingly criticised, particularly by proponents of exploratory data analysis and Bayesian analysis.

Gender: It is a social construction encompassing societal norms and practices of a particular culture. This means that gender practices and identity can differ between and within cultures. Gender differs to sex in that the latter is determined by a person's biology.

Glass ceiling: An invisible glass ceiling that prevents women from progressing in their career. They can see the men in more senior positions through the glass ceiling but cannot easily break through to join them.

Glass escalator: It is a phenomenon whereby males working in a predominantly female dominated occupation, such as nursing, are promoted to management position more quickly that a woman of similar talent.

Inferential statistics: These are statistics generated during inferential analysis that can make generalisations about our research population (the 'N').

Interval variable: It consists of scale interval and ordinal interval data. Scale interval data can be considered as continuous, such as years old, wages earned or the number of friends on Facebook (see **Scale variable**). Ordinal interval data are data such as, measuring an attitude using a 0 to 10 scale where 0 is 'not at all like me' and 10 is 'very much like me'. While there is a numerical structure to an ordinal scale, the data is not quantifiable in the same way as is years old (see **Scale variable**).

Levels of measurement: It is a system for classifying and categorising variables developed by Stevens. Stevens proposed four different types of variables: nominal, ordinal, interval, and ratio.

Mean: The mean is the arithmetic average that is calculated by adding all the values together and dividing the total by the number of observations. One issue to consider when using the mean is how it is susceptible to extreme scores that could give a false impression to the spread of the values.

Measurement: Measurement in the social sciences is used to describe and explain facts of the social world. The use of variables help when measuring such facts (see Levels of Measurement). However, measurement has a good degree of cultural baggage meaning that whatever is measured will always be limited to our current understanding, so can always be improved as our understanding of the social world improves.

Measures of central tendency: It is the value that identifies the central position or middle of our data; often called the average. There are three main ways to calculate measures of central tendency: the mean, the median and the mode. Each approach provides a centre score and is useful for summarising interval (scale or ordinal) data.

Measures of dispersion: It describes the spread of the data. There are several ways to calculate measures of dispersion, including range and standard deviation. They are useful in determining how much variation is in the data and how it is spread out around the average.

Median: The median is the middle value when arranged in an ascending or descending order, meaning that half of the values would be below the median score, and the other half above it. If there is an equal number of scores, then there will be no middle number. In this instance, the two middle numbers are added together and divided by 2. Unlike the mean average, the median average is not susceptible to extreme scores, so it can be used when extreme values are present.

Mode: The mode is the most frequently occurring value in the data's distribution. As there can be more than one most frequently occurring value, the mode can also be considered as bimodal (when there are two frequently occurring values) or multimodal (when there are more than two frequently occurring values).

Narrative: Using a narrative approach in quantitative analysis is the act of developing a story that uses both social science theory and statistical findings.

Nominal variable: A nominal variable is one that names a group such as male/female (gender), younger (under 19 years old) versus older (over 20 years old) (age group); or brother/sister (family member). They differ from ordinal variables in that there is no logical hierarchy to them. You can also only exist in just one of the categories. For instance, you cannot be both in the younger and older age group as your age determines the category to which you belong.

Open data: It is data made available for anyone to use and reuse without any special permissions. In quantitative methodologies, this is often open source data sets.

Ordinal variable: An ordinal variable is unique in that it can exist as either a categorical (ordinal) variable or an interval (ordinal) variable. Its distinguishing feature is how they can be in ranked order where the numbers assigned to the categories reflect the ranking such as age groups (young, middle age and older people). The ranking could also relate to an attitude to something such as when using a Likert scale (Strongly agree to Strongly disagree). However, unlike scale variables, there is no quantifiable distance such as found in years old or the number of friends on Facebook.

Outliers: Outliers are extreme low or high values in the data that can negatively impact on the mean average, making it unrepresentative.

Percent: It literally means 'out of 100' and is a standardised way to express and compare frequencies of categorical data. For example, if a large data set had 4353 men and 5696 women, expressing this as a percentage (43% men and 57% women) would make it easier to compare. Similarly, if a smaller data set had 842 respondents made up of 207 men and 275 women, this would again equate to 43% men and 57% women.

Population: All the members of a specific group that we want to examine, such as the entire population of a country or all the students at a specific university.

Primary data: This is data that is collected by you as the researcher.

Range: The range is the difference between the highest and the lowest value. It is useful in helping summarise the data and showing the degree of spread in the scores.

Range rule: It is an easy way to estimate the standard deviation from the range. To calculate this, divide the range by 4. The range rule should be used with caution.

Raw data: This is data that has not been changed, cleaned or amended prior to use.

Research data: This is data collected for academic research purposes.

Sample: It is part of the population (the 'N') that we are interested in researching. Typically, it is not practical to sample an entire population; so researchers select representative samples (the 'n'). For example, we might take a sample of 40 students out of a class population of 250.

Sampling: This is the process whereby researchers generate their sample from a research population. There are a range of different sampling techniques.

Scale variable: A scale variable uses actual numbers as its source that also has a quantifiable distance between each category. For instance, someone who has 1000 friends on Facebook has 400 more than someone who has 600 friends. Here, the quality of the friendship is not being measured but, rather, the actual number. Another example would be years. If someone is 11 years old, they are 1 year older than someone who is 10 years old. Similarly, someone who is 20 years old is 1 year younger than someone who is 21 years old. The measurement of a year (365/6 days, 52 weeks and 12 months) is considered a constant measure regardless of whether or not there is 365 or 366 days in that particular year.

Secondary data: It is data that someone else has collected and made available for other researchers to use.

Standard deviation: It is a standardised approach to measuring the spread of data from the mean average. Typically, a low standard deviation means the data is clustered more tightly around the mean. Data that has a high standard deviation signifies that it has a lot of spread. Unlike the range, standard deviation allows researchers to compare the spread of data from more than one data set. Standard deviation is also used in assessing different distributions of data.

Statistical narrative: It is the emerging story that is told using numerical data and statistical analysis. In social science research, the narrative should include the social context and appropriate social theory to help build a compelling story that is based on the statistical evidence.

Sticky floor: Low paid low skilled job with little chance of career progression predominantly performed by women.

Univariate analysis: It is analysis using just one variable. Univariate analysis can be either descriptive or inferential statistics. It is used to summarise key information about a variable, such as the numbers of men and women, average age or breakdown of ethnic groups.

Valid percent: It equates to the actual number of responses to a specific question in a data set rather than the number of potential responses that can include missing data. Using the valid percent is a more accurate way to describe the data when using categorical data.

Variable: It is a characteristic or attribute used to measure things in the social world, such as gender, age or attitudes. There are different types of variables, including categorical (nominal and ordinal) and interval (ordinal and scale).

Visualising data: It is the use of graphs, charts, infographics and other visualisations to present, summarise and analyse data.

Check out the next title in the collection: *Statistical Inference and Probability*, **for guidance on Inferential Statistics and Probability.**

REFERENCES

Allen, M. (2017). *The SAGE encyclopedia of communication research methods* (Vols. 1–4). Sage. https://doi.org/10.4135/9781483381411

Ballinger, C. (2011). Why inferential statistics are inappropriate for development studies and how the same data can be better used. *SSRN Electronic Journal*. https://doi.org/10.2139/ssrn.1775002

Barnard, H., Kumar, A., Wenham, A., Smith, E., Drake, B., Collingwood, A., & Leese, D. (2017). *UK poverty 2017*. New York: JRF. Available at: www.jrf.org.uk/report/ukpoverty-2017

Berheide, C. (1992). "Women Still 'Stuck' in Low-Level Jobs." Women in Public Service: A Bulletin of the Center for Women in Government. Albany (NY): Center for Women in Government, SUNY.

Berheide, C. W. (2013). Sticky floor. In V. Smith (Ed.), *Sociology of work: An encyclopedia* (Vol. 1, pp. 826–827). Thousand Oaks, CA: SAGE Publications, Inc. doi: 10.4135/9781452276199.n292

Carvalho, J., Gomes, A. Q., Laja, P., Oliveira, C., Vilarinho, S., Janssen, E., & Nobre, P. (2013). Gender differences in sexual arousal and affective responses to erotica: The effects of type of film and fantasy instructions. *Archives of Sexual Behavior, 42*(6), 1011–1019. https://doi-org.mmu.idm.oclc.org/10.1007/s10508-013-0076-2

Connell, C. (2013). Glass escalator. In V. Smith (Ed.), *Sociology of work: An encyclopedia* (Vol. 1, pp. 332–333). Thousand Oaks, CA: SAGE Publications, Inc. doi: 10.4135/9781452276199.n126

Daniels, D. (2013). Glass ceiling. In V. Smith (Ed.), *Sociology of work: An encyclopedia* (Vol. 1, pp. 330–331). Thousand Oaks, CA: SAGE Publications, Inc. doi: 10.4135/9781452276199.n125

Davies, M. (2015, March 31). Average child will receive £56 worth of Easter eggs and consume 8000 calories this weekend. *Daily Mail Online*. http://www.dailymail.co.uk/health/article-3020028/Average-child-receive-56-worth-Easter-eggs-consume-8-000-calories-weekend.html

Einon, D., (1994). Are Men More Promiscuous Than Women? *Ethology and Sociobiology, 15*(3), pp. 131–143.

Fausto-Sterling, A. (2000). *Sexing the body: Gender politics and the construction of sexuality.* Basic Books.

Fisher, A. D., Ristori, J., Fanni, E., Castellini, G., Forti, G., & Maggi, M. (2016). Gender identity, gender assignment and reassignment in individuals with disorders of sex development: A major of dilemma. *Journal of Endocrinological Investigation, 39*(11), 1207–24. https://doi.org/10.1007/s40618-016-0482-0

Frey, B. (2018). *The SAGE encyclopedia of educational research, measurement, and evaluation* (Vols. 1–4). Sage. https://doi.org/10.4135/9781506326139

Full Fact. (2015, August 10). *Pay gap: Do women earn 80p for every £1 earned by men?* https://fullfact.org/economy/pay-gap-do-women-earn-80p-every-1-earned-men/

Ghaznavi, C., Sakamoto, H., Yoneoka, D. et al. (2019). Trends in heterosexual inexperience among young adults in Japan: analysis of national surveys, 1987–2015. *BMC Public Health, 19*, 355 doi:10.1186/s12889-019-6677-5

Goorwich, S. (2015). The average Brit drinks 876 cups of tea a year – enough to fill two bath tubs. *Metro.* http://metro.co.uk/2015/04/18/the-average-brit-drinks-876-cups-of-tea-a-year-enough-to-fill-two-bath-tubs-5155831/

Hawkes, G. (1996). *A sociology of sex and sexuality.* Open University Press.

Hospido, L., Laeven, L., & Lamo, A. (2019). *The gender promotion gap: Evidence from central banking.* https://voxeu.org/article/gender-promotion-gap-evidence-ecb

Independent Press Standards Organisation. (2015). *09324-15 Muslim Engagement and Development (MEND) v The Sun.* https://www.ipso.co.uk/rulings-and-resolution-statements/ruling/?id=09324-15

Ioannidis, J. P. A. (2018). The challenge of reforming nutritional epidemiologic research. *JAMA Journal of the American Medical Association, 320*(10), 969–970. https://doi.org/10.1001/jama.2018.11025

Johnson, A. (2018). *National survey of sexual attitudes and lifestyles, 2010–2012* (Study No. 7799, 2nd ed.). UK Data Service. http://doc.ukdataservice.ac.uk/doc/7799/mrdoc/UKDA/UKDA_Study_7799_Information.htm

Kettrey, H. H. (2016). What's gender got to do with it? Sexual double standards and power in heterosexual college hookups. *Journal of Sex Research, 53*(7), 754–765. https://doi.org/10.1080/00224499.2016.1145181

Lippa, R. A. (2009). Sex differences in sex drive, sociosexuality, and height across 53 nations: Testing evolutionary and social structural theories. *Archives of Sexual Behavior, 38*(5), 631–651. https://doi.org/10.1007/s10508-007-9242-8

Liszewski, A. (2015). World's largest container ship has four football fields of deck space. *GIZMODO.* https://gizmodo.com/worlds-largest-container-ship-has-four-football-fields-1689154889

MacInnes, J. (2019a). Exploratory data analysis. In P. Atkinson, S. Delamont, A. Cernat, J. W. Sakshaug, & R. A. Williams (Eds.). *SAGE research methods foundations*. Sage. https://doi.org/10.4135/9781526421036889602

MacInnes, J. (2019b). *Statistical significance*. Sage.

Mak, G. (2012). *Doubting sex: Inscriptions, bodies and selves in nineteenth-century hermaphrodite case histories*. Manchester University Press.

McCabe, M. P. (2005). Boys want sex, girls want commitment: Does this trade-off still exist? *Sexual and Relationship Therapy, 20*(2), 139–141. https://doi.org/10.1080/14681990500113252

Mitchell, P. W. (2018). The fault in his seeds: Lost notes to the case of bias in Samuel George Morton's cranial race science. *PLOS BIOLOGY, 16*(10), e2007008. https://doi.org/10.1371/journal. pbio.2007008

Mitricheva, E., Kimura, R., Logothetis, N. K., & Noori, H. R. (2019). Neural substrates of sexual arousal are not sex dependent. *Proceedings of the National Academy of Sciences of the USA, 116*(31), 15671–15676. https://doi.org/10.1073/pnas.1904975116

90% of teens unhappy with body shape. (2004, August 13). *Daily Mail Online*. https://www.dailymail.co.uk/news/article-205285/90-teens-unhappy-body-shape.html

Oakley, A. (1972). *Sex, gender and society*. Temple Smith.

Oakley, A. (2016). *Sex, gender and society*. 10.4324/9781315243399. https://www.researchgate.net/publication/333555600_Sex_gender_and_society

Office for National Statistics. (2015a). *Great Britain population estimates*. 1937 to 2014 https://www.ons.gov.uk/peoplepopulationandcommunity/populationandmigration/populationestimates/adhocs/004357greatbritainpopulationestimates1937to2014

Office for National Statistics. (2015c). *Sexual identity, UK: 2015*. https://www.ons.gov.uk/peoplepopulationandcommunity/culturalidentity/sexuality/bulletins/sexualidentityuk/2015

Office for National Statistics. (2017). *Families and households in the UK: 2017*. https://www.ons.gov.uk/peoplepopulationandcommunity/birthsdeathsandmarriages/families/bulletins/familiesandhouseholds/2017

Office for National Statistics. (2017a). *Families and the labour market, England: 2017*. https://www.ons.gov.uk/employmentandlabourmarket/peopleinwork/employmentandemployeetypes/articles/familiesandthelabourmarketengland/2017

Office for National Statistics. (2017b). *Harmonised concepts and questions for social data sources*. https://gss.civilservice.gov.uk/wp-content/uploads/2019/04/Ethnic-Group-June-17.pdf

Office for National Statistics. (2017c). *Household disposable income and inequality in the UK: Financial year ending 2017*. https://www.ons.gov.uk/peoplepopulationandcommunity/personalandhouseholdfinances/incomeandwealth/bulletins/householddisposableincomeandinequality/financialyearending2017

Office for National Statistics. (2018). *Understanding the gender pay gap in the UK*. https://www.ons.gov.uk/employmentandlabourmarket/peopleinwork/earningsandworkinghours/articles/understandingthegenderpaygapintheuk/2018-01-17

Office for National Statistics. (2020). Crime Survey for England and Wales, 2016–2017 (Study No. 8321, 2nd ed.). UK Data Service. http://doc.ukdataservice.ac.uk/doc/8321/mrdoc/UKDA/UKDA_Study_8321_Information.htm

Olsen, W., Gash, V., Kim, S., & Zhang, M. (2018). *The gender pay gap in the UK: Evidence from the UKHLS*. Government Equalities Office.

Sadoff, C. (2019). *Better data can help us close the global gender gap*. https://www.weforum.org/agenda/2019/03/better-data-can-help-close-the-global-gender-gap/

Sanz, V. (2017). No way out of the binary: A critical history of the scientific production of sex. *Signs, 43*(1), 1–27. https://doi.org/10.1086/692517

Scott Jones, J., & Goldring, John E. (2017). 'Telling stories, landing planes and getting them moving'; a holistic approach to developing students' statistical literacy', *Statistics Education Research Journal* (SERJ), (Special Edition on Statistical Literacy).

Stevens, S.S. (1946). On the theory of scales of measurement. *Science, 103*(2684). 677– 680.

Thompson, C. (2006). Race science. *Theory, Culture & Society, 23*(2–3), 547–549. https://doi.org/10.1177/0263276406023002100

Top female NHS doctors earn much less than their male counterparts, BBC investigation shows. (2018, February 16). *Independent*. https://www.independent.co.uk/news/business/news/nhs-female-doctors-senior-earnings-less-men-gender-pay-gap-a8213416.html

Tukey, J. W. (1970). *Exploratory data analysis*. Pearson.

van Hooff, J. (2015). Desires, expectations and the sexual practices of married and cohabiting heterosexual women. *Sociological Research Online, 20*(4), 123–132. https://doi.org/10.5153/sro.3767

White, P., & Gorard, S. (2017). Against inferential statistics: How and why current statistics teaching gets it wrong. *Statistics Education Research Journal, 16*, 55–65.

Williams, C. L. (1992). The glass escalator: Hidden advantages for men in the "female" professions. *Social Problems, 39*(3), 253–267. https://doi.org/10.1525/sp.1992.39.3.03x0034h

World Economic Forum. (2018). *The global gender gap report 2018*. https://www.weforum.org/reports/the-global-gender-gap-report-2018

Wright, B. D. (1997). A history of social science measurement. *Educational Measurement, Issues and Practice, 16*(4), 33–45. https://doi.org/10.1111/j.1745-3992.1997.tb00606.x

Zaikman, Y., & Marks, M. J. (2014). Ambivalent sexism and the sexual double standard. *Sex Roles, 71*(9–10), 333–344. https://doi-org.mmu.idm.oclc.org/10.1007/s11199-014-0417-1

INDEX

3D pie charts 169
5 Cs of women's work 203, 221
administrative data 27, 28–30, 33–4, 35, 54,
    61, 67, 71, 221
age 11–12, 40, 61, 88–9, 110, 123, 162–3,
    176–88
  car crime 105, 152, 154
  gender pay gap 70–1, 72
  measures of dispersion 142, 143
  Natsal 49, 50–2
  pie charts 165–9
  range 142, 143
  salary 70–1, 132
'Analyse' function 107, 108, 111
anxiety 84–5
average 13, 132–56, 211–13
  feedback scores 133–4
  see also mean; measures of central tendency;
    median; mode

bar charts 148, 161, 162, 164, 169–74,
    189–90, 197
baselines 189–91
Berheide, C. 203
bias 27, 30, 62, 64, 74
bivariate analysis 15, 66, 81, 105, 110–15, 221
  MCT 147–8
  MS Excel 124–7
boxplots 148, 182–5, 197–8
British Social Attitudes Survey (BSA)
    27–8, 35, 43
butter and health 213–16

Carvalho, J. 207
categorical variables 10–11, 13, 15, 72, 73–4,
    77, 78, 81–130, 178, 189
  data cleaning 50
  definition 221
  MS Excel 107, 115–27
  scatterplots 186–7
  SPSS 87–98, 105–15
  see also percentages

Cells 113–14
census data 5, 26, 28, 29, 33, 35, 36, 63–4, 95,
    195–6
central tendency see measures of central
    tendency (MCT)
cholera epidemics 192–3
classification 65–6, 78, 203
clustered bar chart 162, 172
clustered boxplots 184–5
cohort studies 27
Columns 111–12, 115
Connell, C. 202
contingency tables 15, 16, 105–7, 110–16,
    125–7, 161, 171–3, 221
continuous data 12–15, 73, 132, 223
counting 65, 66–9, 78, 85, 203, 221
crime mapping 193–5
Crime Survey of England and Wales (CSEW)
    35, 43, 90–2, 107–10, 117–23, 151–6
cross-sectional surveys 35, 36
cross-tabulations see contingency tables
cumulative percent 11, 98
Cuvier, Georges 62

Daniels, D. 202
data cleaning 16–19, 20–1, 36, 50–3,
    54, 108, 222
data collection 23–54
data familiarisation 92, 222
data labelling 34, 50–1, 170
data management 49–50, 54, 222
data mapping 160, 191–6, 198
DataShine 33, 196
data trimming 52
data visualisation 8, 32–3, 148, 156,
    159–98, 226
dependent variables 112–13
descriptive summaries 99–102, 109, 123
dispersion measures see measures of dispersion
    (MoD)
distance 57–9
Domesday Book 61

double-decker bus calculator 58–9
downloading data 35–48, 54

Einon, D. 209
errors 17, 30, 50, 189, 191
ethics 27, 35
ethnicity 13, 17–18, 56, 61, 62–3, 83, 99
Euripides 56–7
European Social Survey 27, 36
Excel *see* MS Excel
explaining 65, 71–2, 78, 203
'exploded' pie charts 168, 169
exploratory data analysis (EDA) 20–1, 92–3,
    128, 160, 222

fake news 217–18, 222
feedback scores 133–4
file formats 35, 42, 47, 49
frequency 85, 100
    distributions 10–11, 17, 20
    polygon 180
    SPSS 108–10
frequency tables 11–12, 87, 90–9, 100–4, 107,
    128, 160, 222
    MS Excel 117–26
    SPSS 109–10
frequentist approach 9–10, 19, 20, 222
Frey, B. 207
'Full Fact' 218
funding, bias 27

garbage in and garbage out 18, 21, 50
gender 10, 13, 61, 65–72, 83, 93, 105–7,
    171–2, 186–7, 200–2, 222
    5 Cs of women's work 203, 221
    glass ceiling/escalator 202, 222, 223
    one sex 202
    pay gap 70–1, 72, 135, 147–8, 150–1,
        202–5, 212
    ratio 66–9, 71
    religion 110–15, 124–6
    roles 184–5
generalisability 8, 25
General Social Survey (GSS) 27, 33, 35, 36–42,
    141, 164–5, 169–76, 182–8
Geographic Information Systems 192
geospatial data 160, 192–6, 198
glass ceiling 202, 222
glass escalator 202, 223
'The Global Gender Gap Report 2018' 205
Goldring, J. E. 218
graphs 159–98
grouped scatterplots 187

Hawkes, G. 202, 209
height 19, 56, 59, 77, 142
histograms 176–82, 183, 189, 197
horizontal bar charts 170–1

Hospido, L. 202
hypothesis testing *see* frequentist approach

Imperial measures 57–8
Independent Press Standards Organisation
    (IPSO) 217
independent variables 112–13
Index of Multiple Deprivation 33
Industrial Revolution 61
international comparisons 204
interquartile range 14, 183, 185
intersex people 205
interval variables 11–15, 17, 53, 72–6, 78,
    82–5, 128, 131–58
    definition 223
    MS Excel 155–6
    SPSS 151–4
    see also ordinal variables; scale variables
Ioannidis, J. P. A. 216

Joseph Rowntree Foundation 134
journalism 200, 210–18, 219

Likert scales 72, 73, 74–5, 84, 132, 172–3, 224
line charts 174–6, 189, 197
Lippa, R. A. 209
Liszewski, A. 58
literature searches 32, 33, 50
longitudinal surveys 35

McCabe, M. P. 207
matrix scatterplots 187–8
mean 13, 14–15, 52, 132, 134, 136–46,
    148–51, 153–6, 223
measurement 55–79, 203, 205, 223
    legacy 60
    levels 72–3, 78, 223
measures of central tendency (MCT) 11,
    13, 14–15, 16, 20, 51–3, 128, 132–56,
    177–8, 223
    see also mean; median; mode
measures of dispersion (MoD) 13–15, 16, 20,
    128, 133, 141–5, 149, 151–7, 223
    see also range; standard deviation
median 13, 20, 52, 132, 134, 136–7, 139–43,
    145–51, 153–5, 156, 183, 223–4
merged data 98–9, 100
metric measures 57–8
missing data 30, 50–2, 94–7, 101–2
Mitchell, P. W. 210
Mitricheva, E. 207
mode 13, 52, 132, 136–7, 139–43, 145–6,
    149–50, 153, 155, 156, 224
MS Excel 49, 81, 127–8
    categorical data 107, 115–27
    MCT & MoD 155–6
    standard deviation 144
MS Word 126

multimodal data 141, 146
Murphy's Law 50

narrative 10, 71, 150–1, 163, 199–219
  cycle 207–8
  definition 224
National Child Measurement Programme
    (NCMP) 19–20
National Survey of Sexual Attitudes and
    Lifestyles (Natsal) 36, 42–9, 50–2, 161–3,
    166–9, 177–81, 200–1, 206–7, 209–10
Nightingale, Florence 166, 168
nominal variables 72, 73–4, 75–8,
    82–3, 87, 128
  definition 224
  merged data 99
  visualisation 164, 169–70, 197
  see also categorical variables
NOMIS website 35
normal distribution 13, 52–3, 178
null hypothesis significance testing (NHST)
    see frequentist approach

Oakley, A. 65, 201–2
observational data 73, 209, 216, 217, 219
Office for National Statistics (ONS) 5, 18,
    32–3, 62–4, 84–5, 135–6, 142, 149, 203
online shopping hours 138–43
Open Access 31, 34
Open Data 26, 28–9, 30–4, 54, 224
Open Science 31
ordinal variables 72–8, 82–5, 87, 96, 128, 132
  definition 224
  visualisation 164, 169–71, 197
  see also categorical variables; interval
    variables
Outlier Labelling Rule 52
outliers 14, 17, 50, 52, 138–9, 141–2, 145–6,
    177, 182, 186, 209, 224

parameters 5
parametric assumptions 9, 17
percentages 78, 85–128, 134, 166, 189,
    211–12, 224
performance evaluation 28
personal data 31
pie charts 164–9, 189, 197
Pivot Table function 107, 116–25
police data 30, 33, 193–5
population pyramid 181–2
populations 5–8, 67, 69, 104, 224
primary data 24–5, 32, 53–4, 224
probability sampling 7

Q:Q plots 148
Quality of Working Life survey 36–7
quartiles 14, 20, 183–5
questionnaires 24–5

race measurement 61, 62–3
random samples 7–8
range 14, 141–3, 156, 225
range rule 144–5, 225
ratio measures 73
'raw' data 10, 16, 17, 48, 50, 87–8, 104, 106,
    113–14, 164, 197, 225
registration forms 29
reliability 25, 26, 27, 59
research data 27–8, 54, 225
restricted data 34–5
rounding up/down 101
Rows 111–12, 115
'rule of thumb' 58

Sadoff, C. 204
safeguarded data 34
samples 5–9, 225
sample size 7, 97–8, 133
sampling strategies 7, 25, 225
Sanz, V. 205
scale variables 15, 73, 75–7, 78, 88–9, 110,
    132, 134, 223
  data cleaning 50
  definition 225
  visualization 156, 165, 176–85, 189, 197
scatterplots 161–2, 164, 185–8, 197–8
Scott Jones, J. 218
search engines 34
secondary data 18, 225
  collection 24, 25–54
  problems 49
service delivery 28
sexual attitudes 207
sexual orientation 18, 27–8, 40–2, 63–4, 99,
    101–2, 174–6
sexual relationships 200–1, 205–7, 208–10
  celibacy 200–1
  lifetime partners 206–7, 209–10
  see also National Survey of Sexual Attitudes
    and Lifestyles (Natsal)
'Show, Compare or Present' rules 100–1
skewness 17, 177
Snow, John 192–3
Social Research Association 218
SPSS 49, 73, 81, 127–8
  bivariate analysis 110–15
  categorical data 87–98, 105–15
  chart builder tool 163
  contingency tables 106
  CSEW dataset 90–2
  data labelling 50–1
  ice cream dataset 87–90
  MCT & MoD 151–4
  standard deviation 144
  univariate analysis 108–10
stacked bar charts 172–3
stacked histogram 178–9

standard deviation 14–15, 52, 141, 143–5,
    154, 156–7, 225
Stevens Levels of Measurement 72–3, 78, 223
sticky floor 203, 225
surveys 7, 24–5, 35, 84, 95
    European Social Survey 27, 36
    national research 27
    Quality of Working Life survey 36–7
    see also British Social Attitudes Survey
        (BSA); General Social Survey (GSS);
        missing data; National Survey of Sexual
        Attitudes and Lifestyles (Natsal)

third person writing 100
trends 27–8, 36, 40
Tukey, John W. 20, 21, 160–1, 163
Twain, Mark 216–17

UK Data Service (UKDS) 34, 43, 48, 107, 117
univariate analysis 15, 66, 81, 105, 108–10,
    127, 132, 226

validity 26, 59, 78
valid percent 95–8, 100, 101, 106, 121–2, 123,
    126, 128, 226
variables 64–78, 226
    Stevens Levels of Measurement
        72–3, 78, 223
    see also categorical variables; interval
        variables; nominal variables;
        ordinal variables; scale variables
variance 14–15, 52
Versioning 49, 50
visualisation *see* data visualisation

weeks measurement 60
weight 19, 56–7, 59, 77
Williams, C. L. 202
winsorized data 52
writing up results
    averages 149–51, 157
    descriptive summaries 99–102, 123
    third person 100